# AIR CONDITIONING

**Learning Objectives:** Understand the principles of air conditioning and the operation of basic air-conditioning systems. Recognize the characteristics and procedures required to install, operate, and maintain air-conditioning systems

Air conditioning is the simultaneous control of temperature, humidity, air movement, and the quality of air in a conditioned space or building. The intended use of the conditioned space is the determining factor for maintaining the temperature, humidity, air movement, and quality of air. Air conditioning is able to provide widely varying atmospheric conditions ranging from conditions necessary for drying telephone cables to that necessary for cotton spinning. Air conditioning can maintain any atmospheric condition regardless of variations in outdoor weather.

This chapter explains the following subjects as they pertain to air conditioning: principles of air conditioning, heat pumps, chilled-water systems, periodic maintenance, cooling towers, troubleshooting, automotive air conditioning, and ductwork.

## PRINCIPLES OF AIR CONDITIONING

**Learning Objective:** Understand the basic principles of temperature, humidity, and air motion in relation to air conditioning.

Air conditioning is the process of conditioning the air in a space to maintain a predetermined temperature-humidity relationship to meet comfort or technical requirements. This warming and cooling of the air is usually referred to as winter and summer air conditioning.

Here, you are introduced to the operating principles of air-conditioning systems, the environmental factors controlled by air conditioning, and their effects on health and comfort. Refrigerative air conditioners and general procedures pertaining to the installation, operation, and maintenance of these systems are examined. Also, the operation and maintenance of the controls used with these systems are explained.

### TEMPERATURE

Temperature, humidity, and air motion are interrelated in their effects on health and comfort. The term given to the net effects of these factors is effective *temperature*. This effective temperature cannot be measured with a single instrument; therefore, a psychrometric chart aids in calculating the effective temperature when given sufficient known conditions relating to air temperatures and velocity.

Research has shown that most persons are comfortable in air where the effective temperature lies within a narrow range. The range of effective temperatures within which most people feel comfortable is called the COMFORT ZONE. Since winter and summer weather conditions are markedly different, the summer zone varies from the winter zone. The specific effective temperature within the zone at which most people feel comfortable is called the COMFORT LINE (fig. 7-1).

### HUMIDITY

Air at a high temperature and saturated with moisture makes us feel uncomfortable. However, with the same temperature and the air fairly dry, we may feel quite comfortable. Dry air, as it passes over the surface of the skin, evaporates the moisture sooner than damp air and, consequently, produces greater cooling effect. However, air may be so dry that it causes us discomfort. Air that is too dry causes the surface of the skin to become dry and irritates the membranes of the respiratory tract.

HUMIDITY is the amount of water vapor in a given volume of air. RELATIVE HUMIDITY is the amount of water vapor in a given amount of air in comparison with the amount of water vapor the air would hold at a temperature if it were saturated. Relative humidity may be remembered as a fraction or percentage of water vapor in the air; that is, DOES HOLD divided by CAN HOLD.

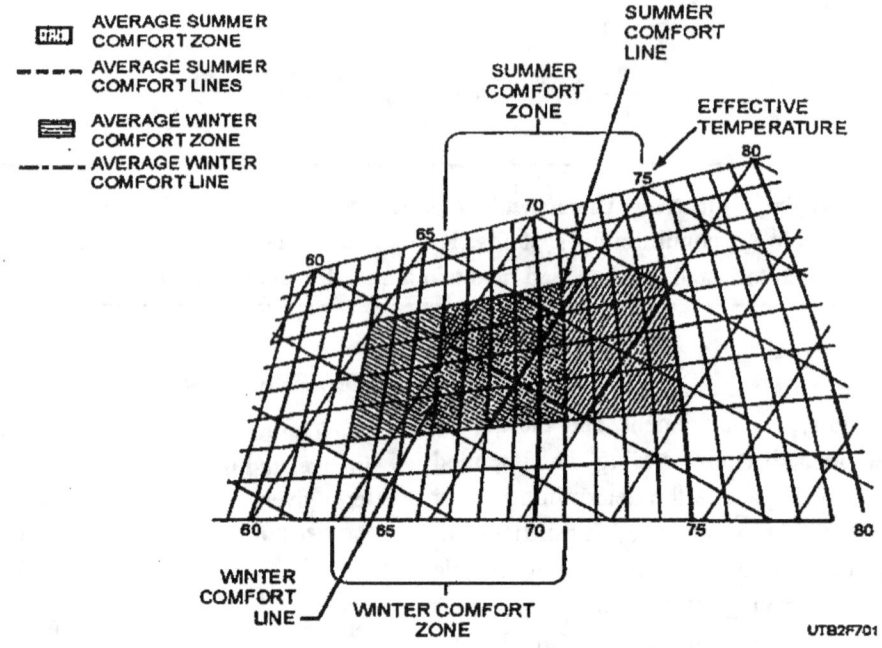

Figure 7-1.—Comfort zones and lines.

Relative humidity is determined by using a sling psychrometer. It consists of a wet-bulb thermometer and a dry-bulb thermometer, as shown in figure 7-2. The wet-bulb thermometer is an ordinary thermometer similar to the dry-bulb thermometer, except that the bulb is enclosed in a wick that is wet with distilled water. The wet bulb is cooled as the moisture evaporates from it while it is being spun through the air. This action causes the wet-bulb thermometer to register a lower temperature than the dry-bulb thermometer. Tables and charts have been designed that use these two temperatures to arrive at a relative humidity for certain conditions.

A comfort zone chart is shown in figure 7-3. The comfort zone is the range of effective temperatures within which the majority of adults feel comfortable. In looking over the chart, note that the comfort zone represents a considerable area. The charts show the wet- and dry-bulb temperature combinations that are comfortable to the majority of adults. The summer comfort zone extends from 66°F effective temperature to 75°F effective temperature for 98 percent of all personnel. The winter comfort zone extends from 63°F effective temperature to 71°F effective temperature for 97 percent of all personnel.

**Dew-Point Temperature**

The dew point depends on the amount of water vapor in the air. If the air at a certain temperature is not

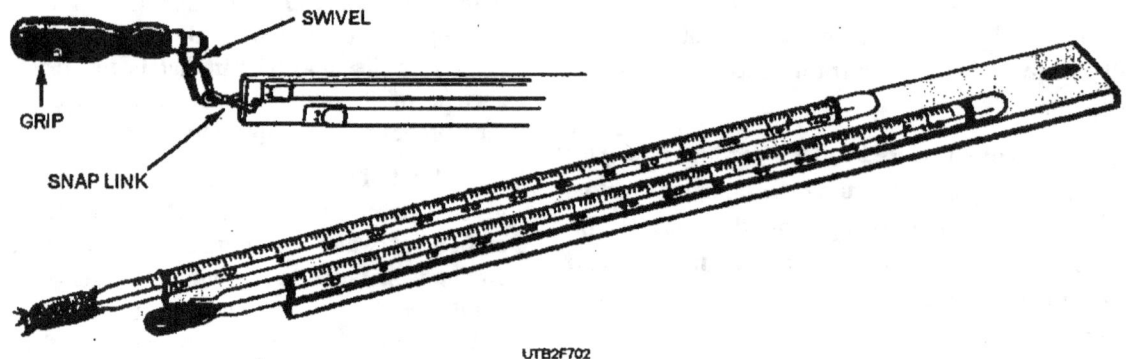

Figure 7-2.—A standard sling psychrometer.

7-2

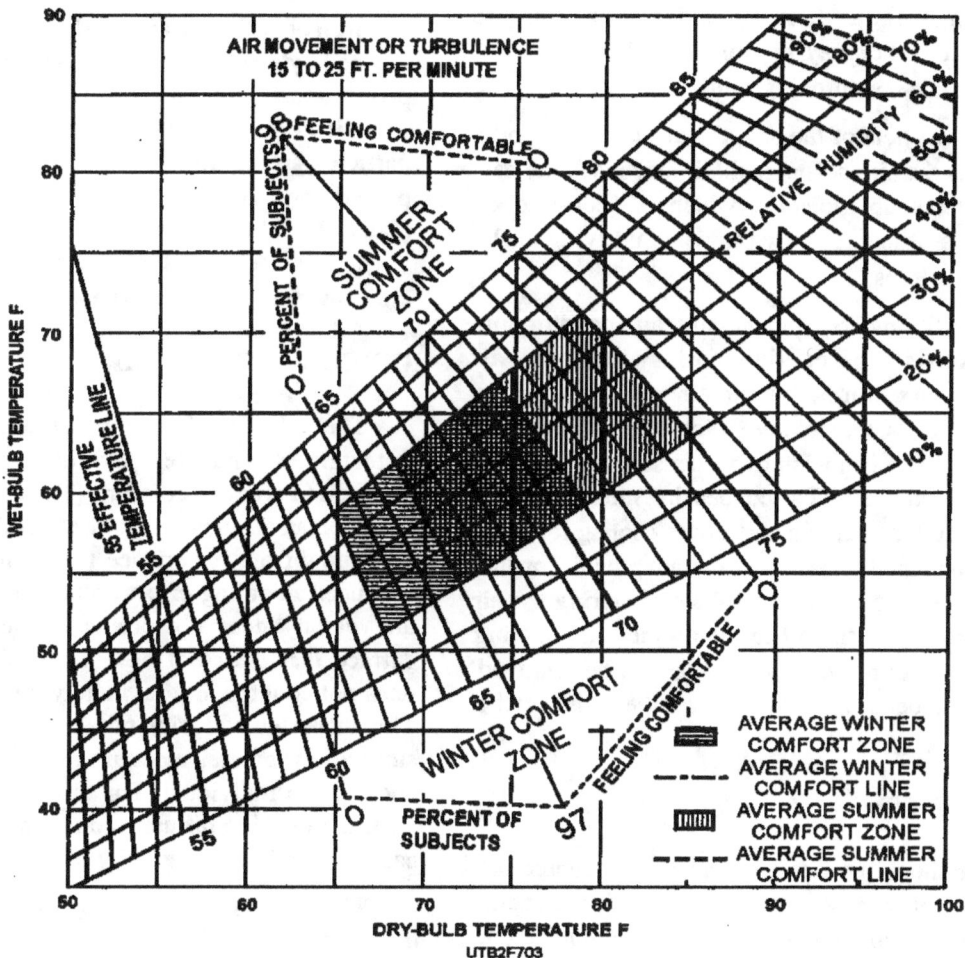

Figure 7-3.—Comfort zone chart.

saturated (maximum water vapor at that temperature) and the temperature of that air falls, a point is finally reached at which the air is saturated for the new and lower temperature, and condensation of the moisture begins. This is the dew-point temperature of the air for the quantity of water vapor present.

### Relationship of Wet-Bulb, Dry-Bulb, and Dew-Point Temperatures

A definite relationship exists between the wet-bulb, dry-bulb, and dew-point temperatures. These relationships are as follows:

- When the air is not saturated but contains some moisture, the dew-point temperature is lower than the dry-bulb temperature, and the wet-bulb temperature is in between.

- As the amount of moisture in the air increases, the amount of evaporation (and, therefore, cooling) decreases. The difference between the temperatures becomes less.

- When the air becomes saturated, all three temperatures are the same and the relative humidity is 100 percent.

To HUMIDIFY air is to increase its water vapor content. To DEHUMIDIFY air is to decrease its water vapor content. The device used to add moisture to the air is a humidifier, and the device used to remove the moisture from the air is a dehumidifier. The control device, sensitive to various degrees of humidity, is called a HUMIDISTAT.

Methods for humidifying air in air-conditioning units usually consist of an arrangement that causes air to pick up moisture. One arrangement consists of a heated water surface over which conditioned air passes and picks up a certain amount of water vapor by evaporation, depending upon the degree of humidifying required. A second arrangement to humidify air is to spray or wash the air as it passes through the air-conditioning unit.

During the heat of the day, the air usually absorbs moisture. As the air cools at night, it may reach the dew point and give up moisture, which is deposited on objects. This principle is used in dehumidifying air by mechanical means.

Dehumidifying equipment for air conditioning usually consists of cooling coils within the air conditioner. As warm, humid air passes over the cooling coils, its temperature drops below the dew point and some of its moisture condenses into water on the surface of the coils. The condensing moisture gives up latent heat that creates a part of the cooling load that must be overcome by the air-conditioning unit. For this reason, the relative humidity of the air entering the air conditioner has a definite bearing on the total cooling load. The amount of water vapor that can be removed from the air depends upon the air over the coils and the temperature of the coils.

## PURITY OF AIR

The air should be free from all foreign materials, such as ordinary dust, rust, animal and vegetable matter, pollen, carbon (soot) from poor combustion, fumes, smoke, and gases. These types of pollution are harmful to the human body alone; however, they include an additional danger because they also carry bacteria and harmful germs. So, the outside air brought into a space or the recirculating air within a space should be filtered during air conditioning.

Air in an air conditioner may be purified or cleaned by filters, air washing, or electricity.

Filters may be designed as permanent or throwaway types. They are usually made of fibrous material, which collects the particles of dust and other foreign matter from the air as it passes through the filter. In some cases, the fibers are dry, while in others they have a viscous (sticky) coating. Filters usually have a large dust-holding capacity. When filters become dust-laden, they are either discarded or cleaned. Permanent filters are usually cleaned. Throwaway filters are only one-time filters and are discarded when they become dust-laden.

Often water sprays are used to recondition the air by washing and cleaning it. These sprays may also serve to humidify or dehumidify the air to some extent.

In some large air-conditioning systems, air is cleaned by electricity. In this type of system, electrical precipitators remove the dust particles from the air. The air is first passed between plates where the dust particles are charged with electricity; then the air is passed through a second set of oppositely charged plates that attract and remove the dust particles (fig. 7-4). This method is by far the best method of air cleaning, but the most expensive.

## CIRCULATION OF AIR

The velocity of the air is the primary factor that determines what temperature and humidity are required to produce comfort. (The chart in figure 7-3 is based on an air movement of 15 to 25 feet per minute.) We know from experience that a high velocity of air produces a cooling effect on human beings. However, air velocity does not produce a cooling effect on a surface that does not have exposed moisture. A fan does not cool the air, but merely increases its velocity. The increased velocity of air passing over the skin surfaces evaporates moisture at a greater rate; thereby, cooling the individual. For this reason, circulation of air has a decided influence on comfort conditions. Air can be circulated by gravity or mechanical means.

When air is circulated by gravity, the cold, and therefore heavier, air tends to settle to the floor, forcing the warm and lighter air to the ceiling. When the air at the ceiling is cooled by some sort of refrigeration, it will settle to the floor and cause the warm air to rise. The circulation of the air by this method will eventually stop when the temperature of the air at the ceiling is the same as the temperature on the floor.

Air may be circulated by mechanical means by axial or radial fans. When either the axial or radial fan is mounted in an enclosure, it is often called a blower.

Q1. *What is the term given to the net effects of temperature, humidity, and air motion?*

Q2. *The comfort line is the specific effective temperature at which most people feel comfortable. True /False*

Q3. *What is the term for the amount of water vapor in a given volume of air?*

Q4. *What instrument is used to measure relative humidity?*

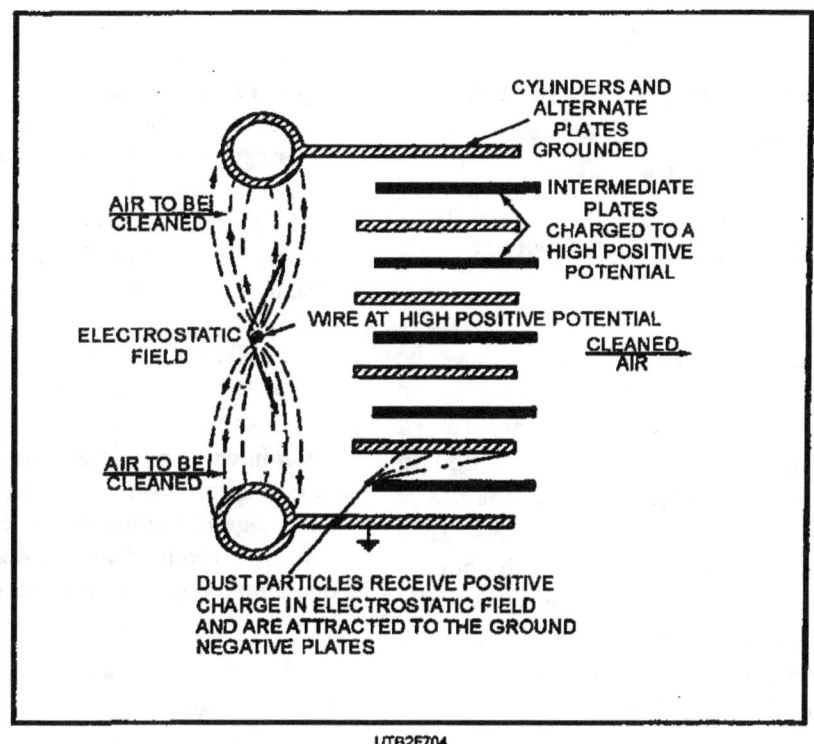

Figure 7-4.—Diagram of an electrostatic filter.

Q5. The point where water vapor condenses is called the dew point. True /False

Q6. What condition exists when the dry-bulb, wet-bulb, and dew-point temperatures are the same?

Q7. What are the two types of filter designs?

Q8. What is the primary factor that determines the temperature and humidity required for room comfort?

## AIR-CONDITIONING SYSTEMS

**Learning Objective:** Recognize basic types of air-conditioning systems, and understand the operation, maintenance, and repair methods and procedures.

A complete air-conditioning system includes a means of refrigeration, one or more heat transfer units, air filters, a means of air distribution, an arrangement for piping the refrigerant and heating medium, and controls to regulate the proper capacity of these components. In addition, the application and design requirements that an air-conditioning system must meet make it necessary to arrange some of these components to condition the air in a certain sequence. For example, an installation that requires re-heating of the conditioned air must be arranged with the re-heating coil on the downstream side of the dehumidifying coil; otherwise, re-heating of the cooled and dehumidified air is impossible.

There has been a tendency by many designers to classify an air-conditioning system by referring to one of its components. For example, the air-conditioning system in a building may include a dual duct arrangement to distribute the conditioned air; therefore, it is then referred to as a dual duct system. This classification makes no reference to the type of refrigeration, the piping arrangement, or the type of controls.

For the purpose of classification, the following definitions are used:

- An air-conditioning unit is understood to consist of a heat transfer surface for heating and cooling, a fan for air circulation, and a means of cleaning the air, motor, drive, and casing.

- A self-contained air-conditioning unit is understood to be an air-conditioning unit that is complete with compressor, condenser, evaporator, controls, and casing.

- An air-handling unit consists of a fan, heat transfer surface, and casing.
- A remote air-handling unit or a remote air-conditioning unit is a unit located outside of the conditioned space that it serves.

## SELF-CONTAINED AIR-CONDITIONING UNITS

Self-contained air-conditioning units may be divided into two types: window-mounted and floor-mounted units. Window-mounted air-conditioning units usually range from 4,000 to 36,000 Btu per hour in capacity (fig. 7-5). The use of windows to install these units is not a necessity. They may be installed in transoms or directly in the outside walls (commonly called a "through-the-wall" installation). A package type of room air conditioner, showing airflow patterns for cooling, ventilating, and exhausting services, is shown in figure 7-6.

In construction and operating principles, the window unit is a small and simplified version of much larger systems. As shown in figures 7-7 and 7-8, the basic refrigeration components are present in the window unit. The outside air cools the condenser coils. The room air is circulated by a fan that blows across the evaporator coils. Moisture, condensed from the humid air by these coils, is collected in a pan at the bottom of the unit; it is usually drained to the back of the unit and discharged. Most window units are equipped with thermostats that maintain a fixed dry-bulb temperature and moisture content in an area within reasonable limits. These units are installed so there is a slight tilt of the unit towards the outside, toward the condenser, to assist in drainage of the condensate. It is a good idea to mount the unit on the eastside of the building to take advantage of the afternoon shade. These units require very little mechanical attention before they are put into operation. Window units are normally operated by the user who should be properly instructed on their use.

Floor-mounted air-conditioning units range in size from 24,000 to 360,000 Btu per hour and are also

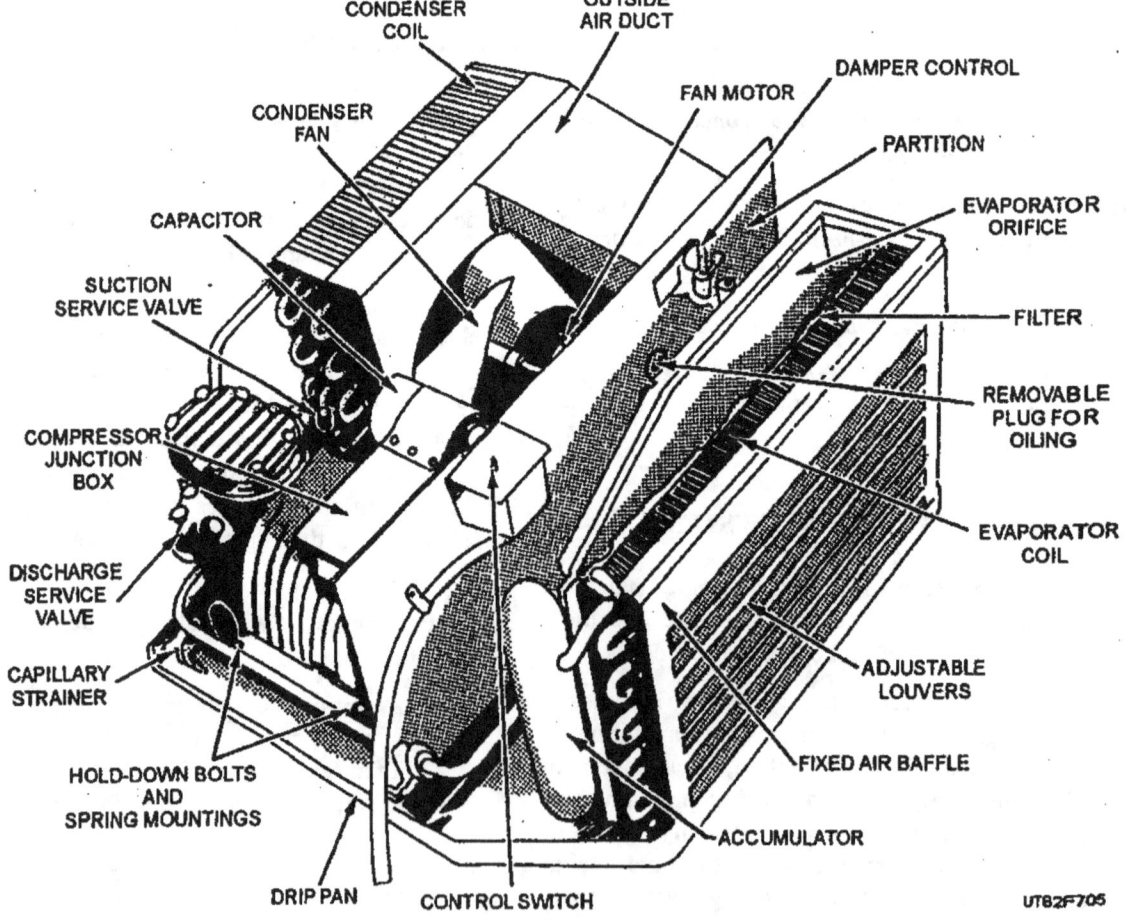

Figure 7-5.—Window air conditioner.

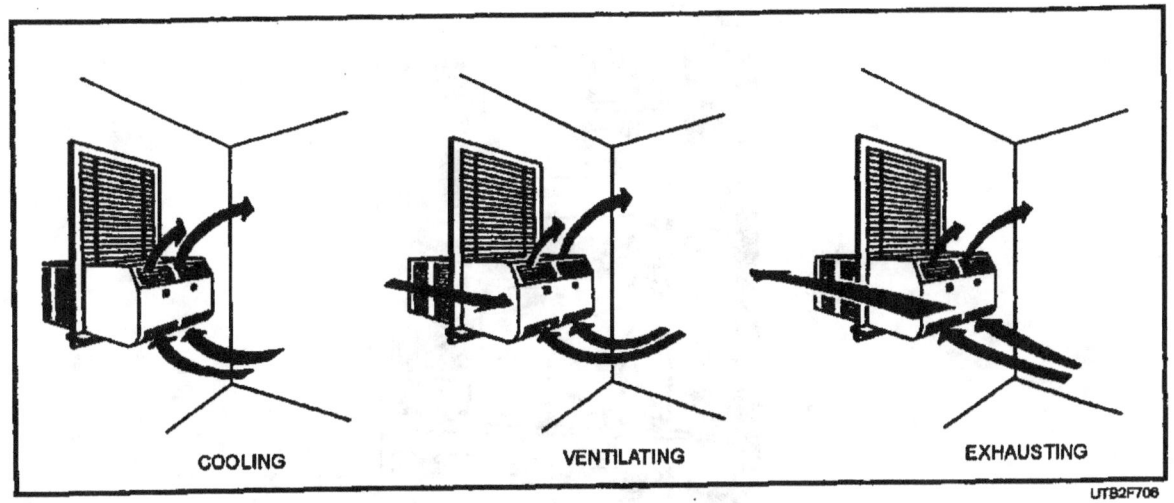

Figure 7-6.—Package type of air-conditioning unit showing airflow patterns.

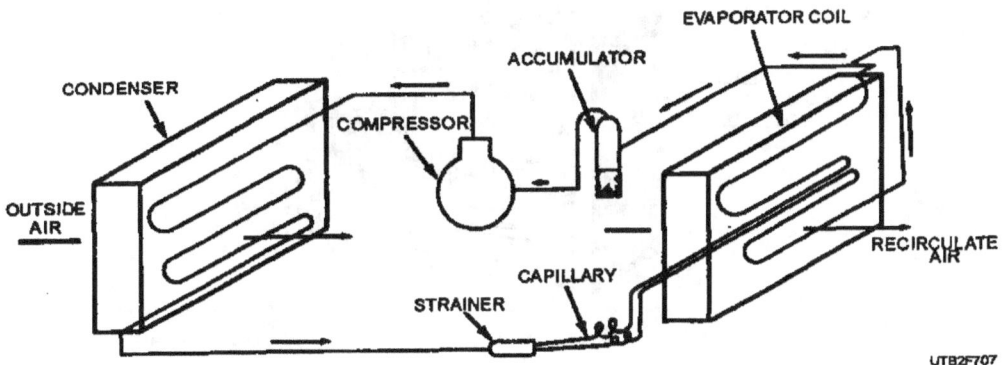

Figure 7-7.—A refrigerant cycle of a package air conditioner.

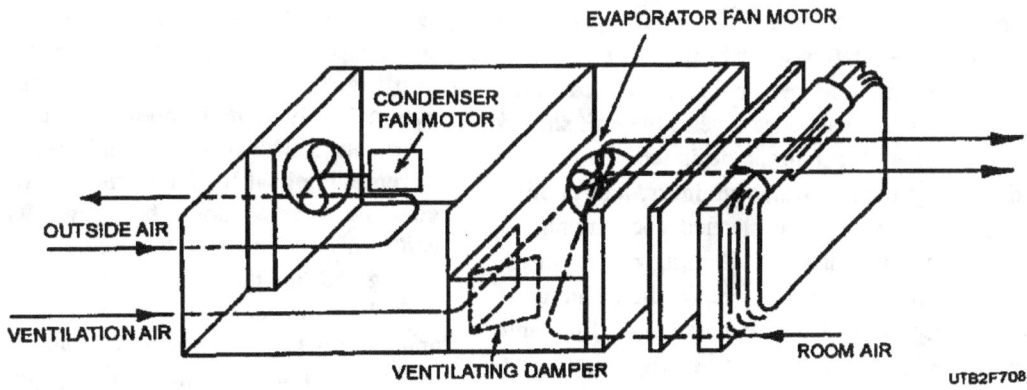

Figure 7-8.—Air-handling components of a package type of room air conditioner.

referred to as PACKAGE units, as the entire system is located in the conditioned space. These larger units, like window units, contain the complete system of refrigeration components. A self-contained unit with panels removed is shown in figure 7-9. These units normally use either a water-cooled or air-cooled condenser.

Self-contained units should be checked regularly to ensure they operate properly. Filters should be

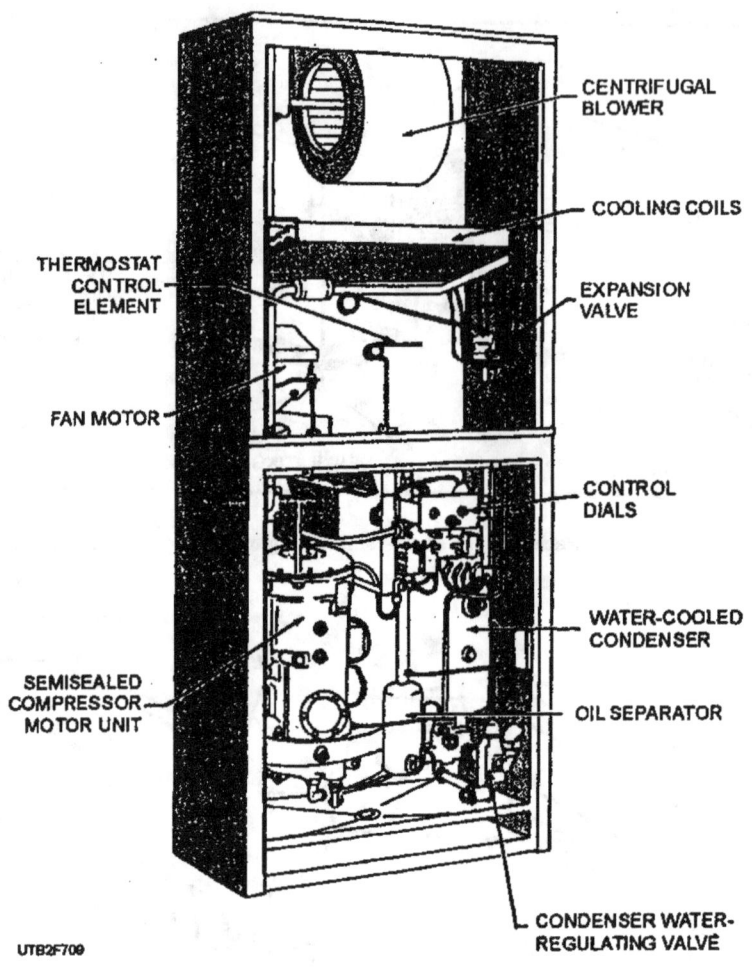

Figure 7-9.—A floor-mounted air-conditioning unit.

renewed or cleaned weekly or more often if necessary. Always stop the blower when changing filters to keep loose dust from circulating through the system. When the filters are permanent, they should be returned to the shop for cleaning. At least once a year, the unit should be serviced. When the unit is designed with spray humidifier, spray nozzle, water strainers, and cooling coils, each device should be cleaned each month to remove water solids and scale. Cooling coil casings, drain pans, fan scrolls, and fan wheels should be wire brushed and repainted when necessary. Oiling and greasing of the blower and motor bearings should be performed as required.

## HEAT PUMPS

A heat pump removes heat from one place and puts it into another. A domestic refrigerator is a heat pump in that it removes heat from inside a box and releases it on the outside. The only difference between a refrigerator and a residential or commercial heat pump is that the latter can reverse its system. The heat pump is one of the most modern means of heating and cooling. Using no fuel, the electric heat pump automatically heats or cools as determined by outside temperature. The air type of unit works on the principle of removing heat from the atmosphere. No matter how cold the weather, some heat can always be extracted and pumped indoors to provide warmth. To cool during the hot months, this cycle is merely reversed with the unit removing heat from the area to be cooled and exhausting it to the outside air. The heat pump is designed to control the moisture in the air and to remove dust and pollen. Cool air, provided during hot weather, enters the area with uncomfortable moisture removed. In winter, when a natural atmosphere is desirable, air is not dried out when pumped indoors.

The heat pump is simple in operation (fig. 7-10). In summer, the evaporator is cooling and the condenser outside is giving off heat the evaporator picked up. In winter, the condenser outside is picking up heat from the outside air because its temperature is lower than that of the outside air (until it reaches the balance point). This heat is then sent to the evaporator by the compressor and is given off into the conditioned space. A reversing valve is the key to this operation. The compressor always pumps in one direction, so the reversing valve changes the hot-gas direction from the condenser to the evaporator as indicated by the setting on the thermostat. The setting of the thermostat assures the operator of a constant temperature through an automatic change from heating to cooling anytime outside conditions warrant. Heat pumps are made not only for small homes but large homes and commercial buildings as well. The heat pump does not require an equipment room, and its minor noise is discharged into the atmosphere. The remote heat pump has only a blower and evaporator, which can be installed under the floor, in an attic, or other out-of-the-way location, depending on the application and its requirements. Supplemental heat can be added into the duct and be set to come on by a second stage of the thermostat, an outside thermostat, or both, depending on design of the system.

### Heating Cycle

The initial heating demand of the thermostat starts the compressor. The reversing valve is de-energized during the heating mode. The compressor pumps the hot refrigerant gas through the indoor coil where heat is released into the indoor air stream. This supply of warmed air is distributed through the conditioned space. As the refrigerant releases its heat, it changes into a liquid, which is then transported to the outdoor coil. The outdoor coil absorbs heat from the air blown across the coil by the outdoor fan. The refrigerant changes from a liquid into a vapor, as it passes through the outdoor coil. The vapor returns to the compressor where it increases temperature and pressure. The hot refrigerant is then pumped back to the indoor coil to start another cycle. A graphic presentation of the nine steps of the cycle is shown in figure 7-11.

### Cooling Cycle

Once the thermostat is put in the cooling mode, the reversing valve is energized. A cooling demand starts the compressor. The compressor pumps hot high-pressure gas to the outdoor coil where heat is released by the outdoor fan. The refrigerant changes into a liquid, which is transported to the indoor blower. The refrigerant absorbs heat from the indoor air of the

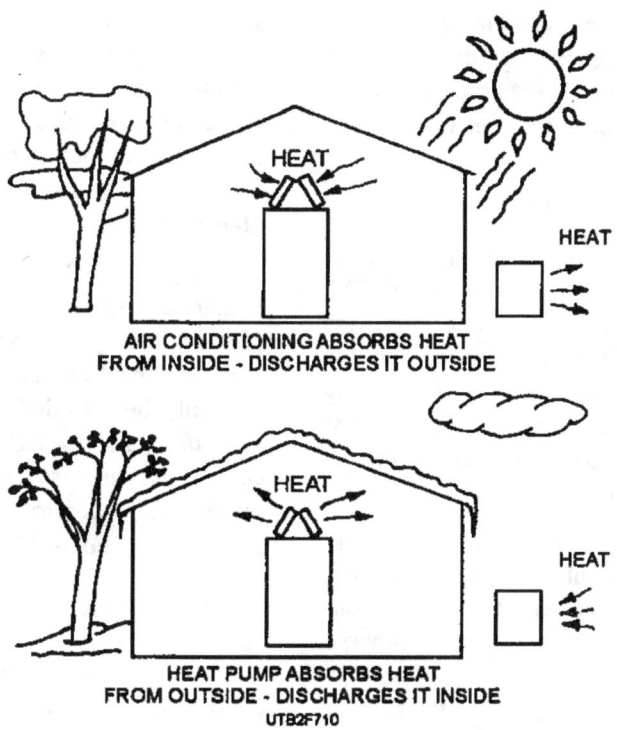

Figure 7-10.—Basic heat pump operation.

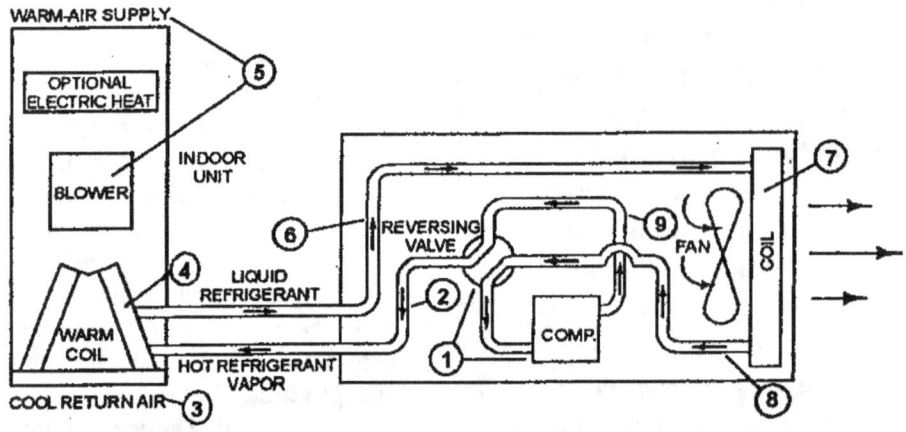

1. THE COMPRESSOR IS ENERGIZED WHILE THE REVERSING VALVE REMAINS DE-ENERGIZED.
2. THE COMPRESSOR PUMPS HOT REFRIGERANT GAS TO THE INDOOR COIL.
3. COOL RETURN AIR IS DRAWN OVER THE INDOOR COIL BY THE BLOWER.
4. THE REFRIGERANT RELEASES ITS HEAT INTO THE RETURN AIRSTREAM AND CONSEQUENTLY TURNS INTO A LIQUID.
5. THE WARMED SUPPLY OF AIR IS DISTRIBUTED THROUGHOUT THE CONTROLLED SPACE.
6. THE LIQUID REFRIGERANT IS TRANSPORTED TO THE OUTDOOR COIL.
7. THE REFRIGERANT ABSORBS HEAT FROM THE OUTDOOR AIR THAT IS BLOWN ACROSS THE COIL BY THE FAN.
8. THE REFRIGERANT TURNS INTO A COOL VAPOR WHICH IS DRAWN BACK TO THE COMPRESSOR.
9. THE COMPRESSOR INCREASES THE PRESSURE OF THE REFRIGERANT. THE HOT REFRIGERANT IS THEN PUMPED BACK TO THE INDOOR COIL TO START ANOTHER CYCLE.

Figure 7-11.—Heating cycle.

supply air, which is distributed throughout the controlled space. This temperature change removes moisture from the air and forms condensate, which must be piped away. The compressor suction pressure draws the cool vapor back into the compressor where the temperature and pressure are greatly increased. This completes the cooling refrigerant cycle. A graphic presentation of the nine steps of the cycle is shown in figure 7-12.

### Defrost Cycle

Heat pumps operating at temperatures below 45°F accumulate frost or ice on the outdoor coil. The relative humidity and ambient temperature affect the degree of accumulation. This ice buildup restricts the airflow through the outdoor coil, which consequently affects the system operating pressures. The defrost control detects this restriction and switches the unit into a defrost mode to melt the ice.

The reversing valve is energized and the machine temporarily goes into the cooling cycle where hot refrigerant flows to the outdoor coil. The outdoor fan stops at the same time, thus allowing the discharge temperature to increase rapidly to shorten the length of the defrost cycle. If there is supplemental heat, a defrost relay activates it to offset the cooling released by the indoor coil.

### Supplemental Heat

As the outside temperature drops, the heat pump runs for longer periods until it eventually operates continually to satisfy the thermostat. The system "balance point" is when the heat pump capacity exactly matches the heating loss. The balance point varies between homes, depending on actual heat loss and the heat pump capacity. However, the balance point usually ranges between 15°F and 40°F. Either electric heat or fossil fuels provide the auxiliary heat.

Conventional heat pump applications use electric heaters downstream from the indoor coil. This design prevents damaging head pressures when the heat pump and auxiliary heat run simultaneously. The indoor coil can only be installed downstream from the auxiliary heat if a "fuelmaster" control system is used. This control package uses a two-stage heat thermostat with the first stage controlling heat pump operation and the second stage controlling furnace operation.

### CHILLED-WATER SYSTEMS

Water chillers (figs. 7-13 and 7-14) are used in air conditioning for large tonnage capacities and for central refrigeration plants serving a number of zones, each with its individual air-cooling and air-circulating

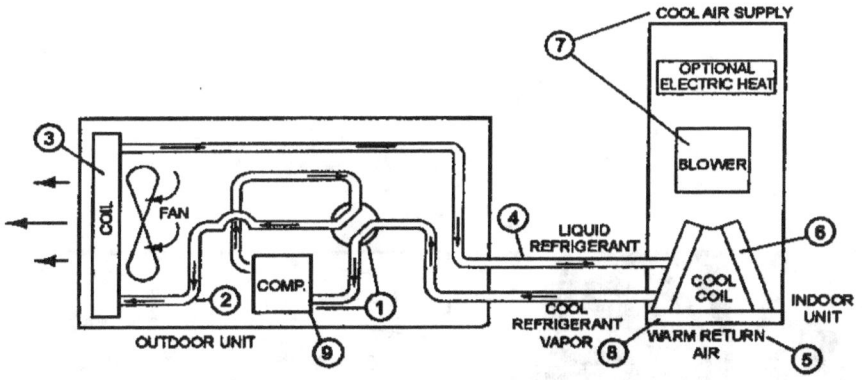

1. THE REVERSING VALVE AND COMPRESSOR ARE ENERGIZED.
2. THE COMPRESSOR PUMPS HOT REFRIGERANT GAS TO THE OUTDOOR COIL.
3. THE FAN DISSIPATES HEAT FROM THE REFRIGERANT AND CHANGES IT INTO A LIQUID.
4. THE LIQUID REFRIGERANT IS SENT ON TO THE INDOOR COIL.
5. WARM AIR IS DRAWN OVER THE INDOOR COIL BY THE BLOWER.
6. THE REFRIGERANT ABSORBS HEAT FROM THE INDOOR AIR AND CHANGES INTO A COOL VAPOR.
7. THIS LOWERS THE TEMPERATURE OF THE SUPPLY AIR WHICH IS DISTRIBUTED THROUGHOUT THE CONTROLLED SPACE.
8. THIS TEMPERATURE CHANGE WILL REMOVE MOISTURE FROM THE AIR AND FORM CONDENSATE WHICH MUST BE PIPED AWAY.
9. THE COMPRESSOR SUCTION PRESSURE DRAWS THE REFRIGERANT BACK INTO THE COMPRESSOR WHERE ITS PRESSURE IS GREATLY INCREASED. THIS COMPLETES ONE COOLING REFRIGERANT CYCLE.

Figure 7-12.—Cooling cycle.

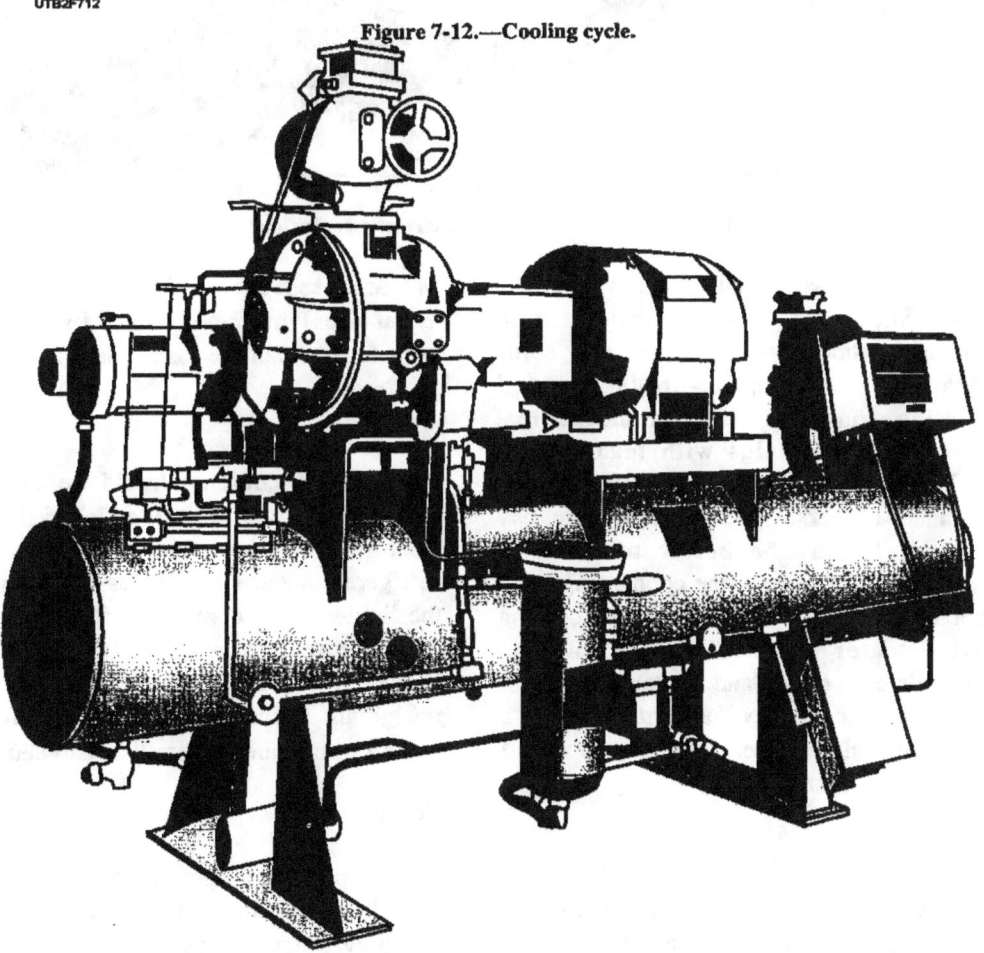

Figure 7-13.—Rotary screw compressor unit.

7-11

Figure 7-14.—Two-stage semihermetic centrifugal unit.

units. An example is a large hospital with wings off a corridor. Air conditioning may be necessary in operating rooms, treatment suites, and possibly some recovery wards. Chilled water-producing and water-circulating equipment is in a mechanical equipment room. Long mains with many joints between condensing equipment and conditioning units increase the chance of leaks. Expensive refrigerant has to be replaced. It may be better to provide water-cooling equipment close to the condensing units and to circulate chilled water to remote air-cooling coils. Chilled water is circulated to various room-located coils by a pump, and the temperature of the air leaving each coil may be controlled by a thermostat that controls a water valve or stops and starts each cooling coil fan motor.

### Types of Coolers

The two most commonly used water coolers (evaporators) for chilled water air conditioning are flooded shell-and-tube and dry-expansion coolers. The disadvantage of the flooded shell-and-tube cooler is that it needs more refrigeration than other systems of equal size. Furthermore, water in tubes may freeze and split tubes when the load falls off.

### Controls

Flooded coolers should be controlled with a low-pressure float control-a float valve placed so the float is about the same level as the predetermined refrigerant level. The float, as a pilot, moves a valve in the liquid line to control the flow of refrigerant to the evaporator. Automatic or thermostatic expansion valves control the dry-expansion coolers. The refrigerant is inside the tubes; therefore, freezing of water on the tubes is less likely to cause damage.

### Condensers

The primary purpose of the condenser is to liquefy the refrigerant vapor. The heat added to the refrigerant in the evaporator and compressor must be transferred to some other medium from the condenser. This medium is the air or water used to cool the condenser.

**WATER-COOLED CONDENSERS.—** Condensing water must be noncorrosive, clean, inexpensive, below a certain maximum temperature, and available in sufficient quantity. The use of corrosive or dirty water results in high maintenance costs for condensers and piping. Dirty water, as from a river, can generally be economically filtered if it is noncorrosive; corrosive water can sometimes be economically treated to neutralize its corrosive properties if it is clean. An inexpensive source of water that must be filtered and chemically treated will probably not be economical to use without some means of conservation, such as an evaporative condenser or a cooling tower.

Water circulated in evaporative condensers and cooling towers must always be treated to reduce the formation of scale, algae, and chalky deposits. Overtreatment of water, however, can waste costly chemicals and result in just as much maintenance as undertreatment.

**SHELL-AND-COIL CONDENSERS.—** A shell-and-coil water-cooled condenser (fig. 7-15) is simply a continuous copper coil mounted inside a steel shell. Water flows through the coil, and the refrigerant vapor from the compressor is discharged inside the shell to condense on the outside of the cold tubes. In many designs, the shell also serves as a liquid receiver.

The shell-and-coil condenser has a low manufacturing cost, but this advantage is offset by the disadvantage that this type of condenser is difficult to service in the field. If a leak develops in the coil, the head from the shell must be removed and the entire coil pulled from the shell to find and repair the leak. A continuous coil is a nuisance to clean, whereas straight tubes are easy to clean with mechanical tube cleaners. In summary, with some types of cooling water, it may be difficult to maintain a high rate of heat transfer with a shell-and-coil condenser.

**SHELL-AND-TUBE CONDENSERS.—** The shell-and-tube water-cooled condenser shown in figure 7-16 permits a large amount of condensing surface to be installed in a comparatively small space. The condenser consists of a large number of 3/4- or 5/8-inch tubes installed inside a steel shell. The water flows inside the tubes while the vapor flows outside around the nest of tubes. The vapor condenses on the outside surface of the tubes and drips to the bottom of the condenser, which may be used as a receiver for the storage of liquid refrigerant. Shell-and-tube condensers are used for practically all water-cooled refrigeration systems.

To obtain a high rate of heat transfer through the surface of a condenser, it is necessary for the water to pass through the tubes at a fairly high velocity. For this reason, the tubes in shell-and-tube condensers are separated into several groups with the same water traveling in series through each of these various groups. A condenser having four groups of tubes is known as a four-pass condenser because the water flows back and forth along its length four times. Four-pass condensers are common although any reasonable number of passes may be used. The fewer the number of water passes in a condenser, the greater the number of tubes in each pass.

The friction of water flowing through a condenser with a few passes is lower than in one having a large number of passes. This means a lower power cost in pumping the water through a condenser with a smaller number of passes.

**TUBE-WITHIN-A-TUBE CONDENSERS.—** The use of tube-within-a-tube for condensing purposes is popular because it is easy to make. Water passing

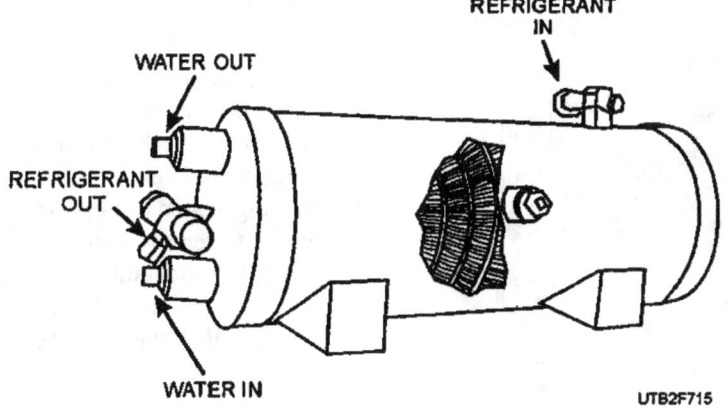

Figure 7-15.—A typical shell-and-coil water-cooled condenser.

7-13

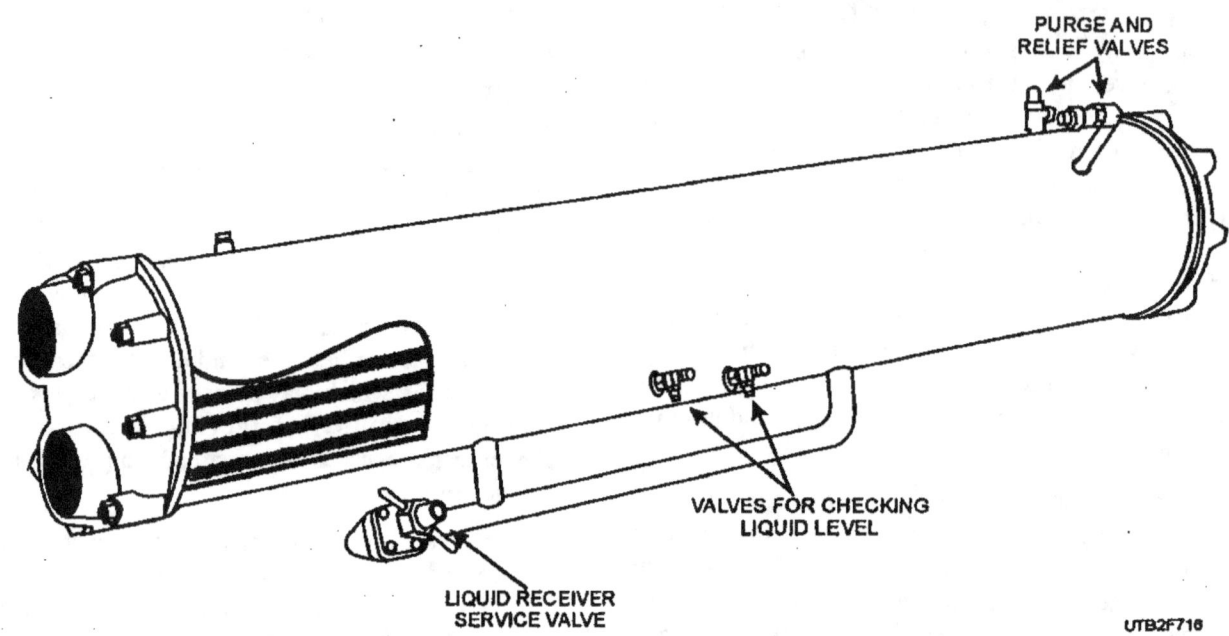

Figure 7-16.—A typical shell-and-tube condenser-liquid receiver.

through the inner tube along with the exterior air condenses (fig 7-17) the refrigerant in the outer tube. This "double cooling" improves efficiency of the condenser. Water enters the condenser at the point where the refrigerant leaves the condenser. It leaves the condenser at the point where the hot vapor from the compressor enters the condenser. This arrangement is called counterflow design.

The rectangular type of tube-within-a-tube condenser uses a straight, hard copper pipe with manifolds on the ends. When the manifolds are removed, the water pipes can be cleaned mechanically.

## CLEANING WATER-COOLED CONDENSERS

You may be assigned to some activities where water-cooled condensers are used in the air-conditioning system. So, the Utilitiesman will probably have the job of cleaning the condensers. Information that assists you in cleaning water-cooled condensers is presented below.

Water contains many impurities—the content of which varies in different localities. Lime and iron are especially injurious; they form a hard scale on the walls of water tubes that reduces the efficiency of the condenser. Condensers can be cleaned mechanically or chemically.

Scale on tube walls of condensers with removable heads is removed by attaching a round steel brush to a rod and by working it in and out of the tubes. After the tubes have been cleaned with a brush, flush them by running water through them. Some scale deposits are harder to remove than others, and a steel brush may not do the job. Several types of tube cleaners for removing hard scale can usually be purchased from local sources. Be sure that the type selected does not injure water tubes.

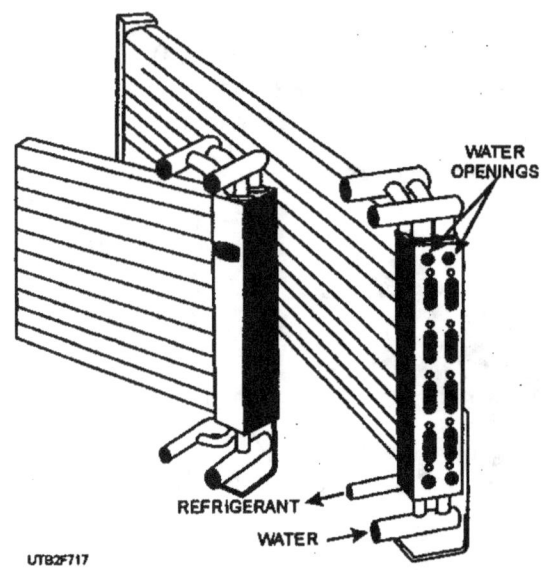

Figure 7-17.—Tube-within-a-tube condenser designed to permit cleaning of water tubes.

The simplest method of removing scale and dirt from condenser tubes not accessible for mechanical

7-14

cleaning is by using inhibited acid to clean coils or tubes through chemical action. Figure 7-18 shows the connections and the equipment for cleaning the condenser with an inhibited acid, both when the acid flows by gravity (view 1) and when forced circulation is used (view 2). When scale deposit is not great, gravity flow of the acid provides enough cleaning. When the deposit almost clogs the tubes, forced circulation should be used.

**WARNING**

Prevent chemical solution from splashing in your eyes and on your skin or clothing.

Equipment and connections for circulating inhibited acid through the condenser using gravity flow, as shown in figure 7-18, view 1, are as follows:

1. A rubber or plastic bucket for mixing solution. Do not use galvanized materials because prolonged contact with acid deteriorates such surfaces.

2. A crock or wooden bucket for catching the drainage residue.

3. One-inch steel pipe that is long enough to make the connections shown.

4. Fittings for 1-inch steel pipe. The vent pipe shown should be installed at the higher connection of the condenser.

Equipment and connections for circulating inhibited acid through the condenser using forced circulation, as shown in figure 7-18, are as follows:

1. A pump suitable for this application. A centrifugal pump and a 1/2-horsepower motor is recommended (30 gallons per minute at 35-foot head capacity).

2. A nongalvanized metal tank, stone or porcelain crock, or wooden barrel with a capacity of about 50 gallons with ordinary bronze or copper screening to keep large pieces of scale or dirt from getting into the pump intakes.

3. One-inch pipe that is long enough to make the piping connections shown.

4. Fittings for 1-inch steel globe valves. The vent pipe, as shown, should be installed at the higher connection of the condenser.

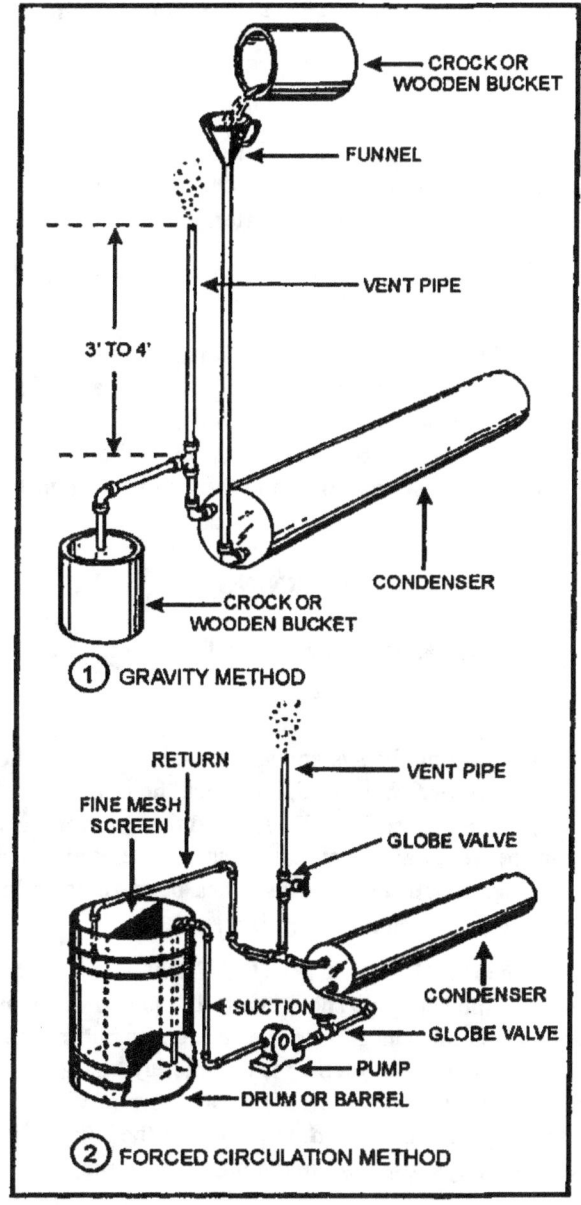

Figure 7-18.—Cleaning water-cooled condensers with acid solution.

Handle the inhibited acid for cleaning condensers with the usual precautions observed when handling acids. It stains hands and clothing and attacks concrete and if an inhibitor is not present, it reacts with steel. Therefore, use every precaution to prevent spilling or splashing. When splashing might occur, cover the surfaces with burlap or boards. Gas produced during cleaning that escapes through the vent pipe is not harmfu but prevents any liquid or spray from being carried hrough with the gas. The basic formula should be maintained as closely as possible, but a variation of 5 percent is permissible. The inhibited acid solution is made up of the following:

1. Water.

2. Commercial hydrochloric (muriatic) acid with specific gravity of 1.19. Eleven quarts of acid should be used for each 10 gallons of water.

3. Three and two-fifths ounces of inhibitor powder for each 10 gallons of water used.

4. Place the required amount of water in a nongalvanized metal tank or wooden barrel, and add the necessary amount of inhibitor powder while stirring the water. Continue stirring the water until the powder is completely dissolved; then add the required quantity of acid.

### WARNING

NEVER add water to acid; this mistake may cause an explosion.

In charging the system with an acid solution when GRAVITY FLOW is used, introduce the inhibited acid as shown in figure 7-18. Do not add the solution faster than the vent can exhaust the gases generated during cleaning. When the condenser has been filled, allow the solution to remain overnight.

When FORCED CIRCULATION is used, the valve in the vent pipe should be fully opened while the solution is introduced into the condenser but must be closed when the condenser is completely charged and the solution is circulated by the pump. When a centrifugal pump is used, the valve in the supply line may be fully closed while the pump is running.

The solution should be allowed to stand or be circulated in the system overnight for cleaning out average scale deposits. The cleaning time also depends on the size of the condenser to be cleaned. For extremely heavy deposits, forced circulation is recommended, and the time should be increased to 24 hours. The solution acts more rapidly if it is warm, but the cleaning action is just as thorough with a cold solution if adequate time is allowed.

After the solution has been allowed to stand or has been circulated for the required time through the condenser, it should be drained and the condenser thoroughly flushed with water. To clean condensers with removable heads by using inhibited acid, use the above procedure without removing the heads.

However, extra precaution must be exercised in flushing out the condenser with clear water after the acid has been circulated through the condenser to ensure acid removal from all water passages.

## MAINTENANCE

A well-planned maintenance program avoids unnecessary downtime, prolongs the life of the unit, and reduces the possibility of costly equipment failure. It is recommended that a maintenance log be maintained for recording the maintenance activities. This action provides a valuable guide and aids in obtaining extended length of service from the unit. This section describes specific maintenance procedures, which must be performed as a part of the maintenance program of the unit. Use and follow the manufacturer's manual for the unit you are to do maintenance on. When specific directions or requirements are furnished, follow them. Before performing any of these operations, however, ensure that power to the unit is disconnected unless otherwise instructed.

### WARNING

When maintenance checks and procedures must be completed with the electrical power on, care must be taken to avoid contact with energized components or moving parts. Failure to exercise caution when working with electrically powered equipment may result in serious injury or death.

### Coil Cleaning

Refrigerant coils must be cleaned at least once a year or more frequently if the unit is located in a dirty environment. This action helps maintain unit operating efficiency and reliability. The relationship between regular coil maintenance and efficient/reliable unit operation is as follows:

- Clean condenser coils minimize compressor head pressure and amperage draw and promote system efficiency.

- Clean evaporator coils minimize water carry-over and helps eliminate frosting and/or compressor flood-back problems.

- Clean coils minimize required fan brake horsepower and maximize efficiency by keeping coil static pressure loss at a minimum.
- Clean coils keep the motor temperature and system pressure within safe operating limits for good reliability.

The following equipment is required to clean condenser coils: a soft brush and either a garden pump-up sprayer or a high-pressure sprayer. In addition, a high-quality detergent must be used. Follow the manufacturer's recommendations for mixing to make sure the detergent is alkaline with a pH value less than 8.5.

Specific steps required for cleaning the condenser coils are as follows:

1. Disconnect the power to the unit.

**WARNING**

Open the unit disconnect switch. Failure to disconnect the unit from the electrical power source may result in severe electrical shock and possible injury or death.

2. Remove enough panels from the unit to gain access to the coil.
3. Protect all electrical devices, such as motors and controllers, from dust and spray.
4. Straighten coil fins with a fin rake, if necessary.
5. Use a soft brush to remove loose dirt and debris from both sides of the coil.
6. Mix the detergent with water according to the manufacturer's instructions. The detergent and water solution may be heated to a maximum of 150°F to improve its cleaning ability.

**WARNING**

Do not heat the detergent and water solution to temperatures in excess of 150°F. High-temperature liquids sprayed on the coil exterior raise the pressure within the coil and may cause it to burst. Should this occur, the result could be both injury to personnel and equipment damage.

7. Place the detergent and water solution in the sprayer. If a high-pressure sprayer is used, be sure to follow these guidelines:
- Minimmum nozzle spray angle is 15 degrees.
- Spray the solution perp endicular (at a 90-degree angle) to the coil face.
- Keep the sprayer nozzle at least 6 inches from the coil.
- Sprayer pressure must not exceed 600 psi.

**CAUTION**

Do NOT spray motors or other electrical components. Moisture from the spray can cause component failure.

8. Spray the side of the coil where the air leaves first; then, spray the other side (where the air enters). Allow the detergent and water solution to stand on the coil for 5 minutes.
9. Rinse both sides of the coil with cool water.
10. Inspect the coil and if it still appears dirty, repeat Steps 8 and 9.
11. Remove the protective covers installed in Step 3.
2. Replace all unit panels and parts, and restore electrical power to the unit.

**Fan Motors**

Inspect periodically for excessive vibration or temperature. Operating conditions vary the frequency of inspection and lubrication. Motor lubrication instructions are found on the motor tag or nameplate. If not available, contact the motor manufacturer for instructions.

To re-lubricate the motor, complete the following:

**WARNING**

Disconnect the power source for motor lubrication. Failure to do so may result in injury or death from electrical shock or moving parts.

1. Turn the motor off. Make sure it cannot accidentally restart.

2. Remove the relief plug and clean out any hardened grease.

3. Add fresh grease through the fitting with a low-pressure grease gun.

4. Run the motor for a few minutes to expel any excess grease through the relief vent.

5. Stop the motor and replace the relief plug

**Fan Bearing Lubrication**

Fan bearings with grease fittings or with grease line extensions should be lubricated with a lithium-base grease that is free of chemical impurities. Improper lubrication can result in early bearing failure. To lubricate the fan bearings, complete the following:

1. Lubricate the bearings while the unit is not running; disconnect the main power switch.

2. Connect a manual grease gun to the grease line or fitting.

3. Add grease, preferably when the bearing is warm, while turning the fan wheel manually until a light bead of grease appears at the bearing grease seal.

**Filters**

To clean permanent filters, wash under a stream of hot water to remove dirt and lint. Follow with a wash of mild alkali solution to remove old filter oil. Rinse thoroughly and let dry. Recoat both sides of the filter with filter oil and let dry. Replace the filter element in the unit.

### CAUTION

Always install filters with directional arrows pointing toward the fans.

**PERIODIC MAINTENANCE**

Perform all of the indicated maintenance procedures at the intervals scheduled. This prolongs the life of the unit and reduces the possibility of costly equipment failure and downtime. A checklist should be prepared which lists the required service operations and the times at which they are to be performed. The following is a sample of such a list.

**Weekly**

1. Check the compressor oil level. If low, allow the compressor to operate continually at full load for 3 to 4 hours; check the oil level at 30-minute intervals. If the level remains low, add oil.

2. Observe the oil pressure. The oil pressure gauge reading should be approximately 20 to 35 psi above the suction pressure gauge reading.

3. Stop the compressor and check the shaft seal for excessive oil leakage. If found, check the seal with a refrigerant leak detector (open compressor only).

4. Check the condition of the air filters and air-handling equipment. Clean or replace filters, as necessary.

5. Check the general operating conditions, system pressures, refrigerant sight glass, and so forth.

**Monthly**

(Repeat Items 1 through 5)

6. Lubricate the fan and motor bearings, as necessary. Obtain and follow the manufacturer's lubricant specifications and bearing care instructions.

7. Check the fan belt tension and alignment.

8. Tighten all fan sheaves and pulleys. If found to be loose, check alignment before tightening.

9. Check the condition of the condensing equipment. Observe the condition of the condenser coil in the air-cooled condenser. Clean, as necessary. Check the cooling tower water in the water-cooled condenser. If algae or scaling is evident, water treatment is needed. Clean the sump strainer screen of the cooling tower.

**Annually**

(Repeat Items 1 through 9)

10. Drain all circuits of the water-condensing system. Inspect the condenser piping and clean any scale or sludge from the tubes of the condenser.

11. If a cooling tower or evaporative condenser is used, flush the pumps and sump tank. Remove any rust or corrosion from the metal surfaces and repaint.

12. Inspect all motor and fan shaft bearings for signs of wear. Check the shafts for proper end-play adjustment.

13. Replace worn or frayed fan belts.

14. Clean all water strainers.

15. Check the condition of the ductwork.

16. Check the condition of the electrical contacts of all contactors, starters, and controls. Remove the condensing unit control box cover and inspect the panel wiring. All electrical connections should be secure. Inspect the compressor and condenser fan motor contactors. If the contacts appear severely burned or pitted, replace the contactor. Do not clean the contacts. Inspect the condenser fan capacitors for visible damage.

### Seasonal Shutdown

In preparation for seasonal shutdown, it is advisable to pump down the system and valve off the bulk of the refrigerant charge in the condenser. This action minimizes the quantity of refrigerant that might be lost due to any minor leak on the low-pressure side of the system, and, in the case of the open compressor, refrigerant that might leak through the shaft seal.

The following steps should be followed for the hermetic compressor pump down.

1. Close the liquid line shutoff valve at the condenser and start the system. When the suction pressure drops to the cutout setting of the low-pressure control, the compressor stops.

2. Open the compressor electrical disconnect switch to prevent the compressor from restarting, and then front-seat the compressor discharge and suction valves.

The following steps should be followed for the open compressor pump down.

1. If the system is not equipped with gauges, install a pressure gauge in the back-seat port of the compressor suction valve. Crack the valve off the backseat.

2. Close the liquid line shutoff valve at the condenser.

3. Manually open the liquid line solenoid valve(s). If the valves do not have manual opening devices, lower the setting of the system temperature controller so the valves are held open during the pump down.

4. Install a jumper wire across the terminals of the low-pressure switch. Since the system suction pressure is to be pumped down below the cutout setting of the low-pressure switch, the jumper is necessary to keep the compressor running.

5. Start the compressor. Watching the suction pressure gauge, stop the compressor by opening its electrical disconnect switch when the gauge reading reaches 2 psig.

6. Front-seat the compressor discharge valve.

### CAUTION

Do not allow the compressor to pump the suction pressure into a vacuum. A slight positive pressure is necessary to prevent air and moisture from being drawn into the system through minor leaks and through the now unmoving shaft seal.

7. Remove the jumper wire from the low-pressure control.

8. Remove the gauge from the port of the suction valve; replace the port plug and front-seat the valve.

The following steps are required for all systems:

1. Using a refrigerant leak detector, check the condenser and liquid receiver, if used, for refrigerant leaks.

2. Valve off the supply and return water connections of the water-cooled condenser. Allow the condenser to remain full of water during the off season. A drained condenser shell is more likely to rust and corrode than one full of water. If the condenser will be subjected to freezing temperatures, drain the water and refill it with an antifreeze solution.

3. Drain the cooling tower or evaporative condenser, if used; flush the sump and paint any rusted or corroded areas.

4. Open the system master disconnect switch and padlock it in the OPEN position.

### Seasonal Start-up

The steps to follow for the seasonal start-up are as follows:

1 Perform all annual maintenance on the air-handling system and other related equipment.

2 Fill the water sump of the cooling tower or evaporative condenser, if used.

3. Open the shutoff valves of the water-cooled condenser.

4. Make certain the liquid line solenoid valve(s) is on automatic control.

5. Open the liquid line shutoff valve.

6. Back-seat the compressor suction and discharge valves.

7. Close the system master electrical disconnect switch.

8. Start the system.

9. After the system has operated for 15 to 20 minutes, check the compressor oil level sight glass, oil pressure, and the liquid line sight glass. If satisfactory, readjust the system temperature controller to the proper temperature setting.

## SAFETY WARNINGS

Most units used for comfort air conditioning operate using R-12 or R-22 refrigerants that are not toxic except when decomposed by a flame. If the liquefied refrigerant contacts the eyes, the person suffering the injury must be taken to a doctor at once.

Should the skin come in contact with the liquefied refrigerant, the skin is to be treated as though it had been frostbitten or frozen. Refer to NAVEDTRA 13119, *Standard First Aid for Treatment of Frostbite.*

Do not adjust, clean, lubricate, or service any parts of equipment that are in motion. Ensure that moving parts, such as pulleys, belts, or flywheels, are fully enclosed with proper guards attached.

Before making repairs, open all electric switches controlling the equipment. Tag and lock the switches to prevent short circuits or accidental starting of equipment. When moisture and brine are on the floor, fatal grounding through the body is possible when exposed electrical connections can be reached or touched by personnel. De-energize electrical lines before repairing them, and ground all electrical tools.

Q9. What are the two types of self-contained air-conditioning units?

Q10. Who normally operates a window air-conditioning unit?

Q11. Floor units are often referred to as what type of unit?

Q12. What type of unit cools and heats by reversing its cycle?

Q13. The reversing valve changes the direction of hot gas from what component to what other component in the system?

Q14. Heat pumps operating below 45°F develop what problem on the outside coils?

Q15. When the capacity of a heat pump matches the heat loss, it has reached what point?

Q16. What are the two most common evaporators used for water chiller systems?

Q17. Dry-expansion evaporators are controlled by what type of expansion valve?

Q18. Why is a tube-within-a-tube condenser popular?

Q19. Condensers can be cleaned in what two ways?

Q20. Refrigerant coils should be cleaned at least how often?

## MAJOR SYSTEM COMPONENTS AND CONTROLS

**Learning Objectives:** Recognize and understand different types of cooling towers, compressors, and controls. Understand basic maintenance requirements for cooling towers.

In this section, cooling towers and compressors which are the two major components of an air-conditioning system are discussed. In addition, the major control elements of an air-conditioned system are also covered.

## COOLING TOWERS

Cooling towers are classified according to the method of moving air through the tower as natural draft, induced draft, or forced draft (figs. 7-19 and 7-20).

### Natural Draft

The natural draft cooling tower is designed to cool water by means of air moving through the tower at the low velocities prevalent in open spaces during the summer. Natural draft towers are constructed of cypress or redwood and have numerous wooden decks of splash bars installed at regular intervals from the bottom to the top. Warm water from the condenser is

Figure 7-19.—A package tower with a remote, variable speed pump.

Figure 7-20.—Paralleled package towers.

flooded or sprayed over the distributing deck and flows by gravity to the water-collecting basin.

A completely open space is required for the natural draft tower since its performance depends on existing air currents. Ordinarily, a roof is an excellent location. Louvers must be placed on all sides of a natural draft tower to reduce drift loss.

Important design considerations are the wind velocity and the height of the tower. A wind velocity of 3-miles per hour is generally used for a design of natural draft cooling towers. The natural draft cooling tower was once the standard design for cooling condenser water in refrigeration systems up to about 75 tons. It is now rarely selected unless low initial cost and minimum power requirements are primary considerations. The drift loss and space requirements are much greater than for other cooling tower designs.

### Induced Draft

An induced draft cooling tower is provided with a top-mounted fan that induces atmospheric air to flow up through the tower, as warm water falls downward. An induced draft tower may have only spray nozzles for water backup, or it may be filled with various slat and deck arrangements. There are several types of induced draft cooling towers.

In a counterflow induced draft tower (fig. 7-21, C), a top-mounted fan induces air to enter through the bottom of the tower and to flow vertically upward as the water cascades down through the tower. The counterflow tower is particularly well adapted to a restricted space as the discharge air is directed vertically upward, and if equipped with a inlet on each side, requires only minimum clearance for air intake area. The primary breakup of water may be either by pressure spray or by gravity from pressure-filled flumes.

A parallel-flow induced draft tower (fig. 7-21, A) operates the same way as a counter-flow tower, except the top-mounted fan pulls the air in through the top of the tower and pushes it out the bottom. The airflow goes in the same direction as the water.

Comparing counterflow and parallel-flow induced draft towers of equal capacity, the parallel-flow tower is somewhat wider but the height is much less. Cooling towers must be braced against the wind. From a structural standpoint, therefore, it is much easier to design a parallel-flow than a counterflow tower, as the low silhouette of the parallel-flow type offers much less resistance to the force of the winds.

Mechanical equipment for counterflow and parallel-flow towers is mounted on top of the tower and is readily accessible for inspection and maintenance. The water-distributing systems are completely open on top of the tower and can be inspected during operation. This makes it possible to adjust the float valves and clean stopped-up nozzles while the towers are operating.

The cross-flow induced draft tower (fig. 7-21, B) is a modified version of the parallel-flow induced draft tower. The fan in a cross-flow cooling tower draws air through a single horizontal opening at one end and discharges the air at the opposite end.

The cooling tower is a packaged tower that is inexpensive to manufacture and is extremely popular for small installations. As a packaged cooling tower

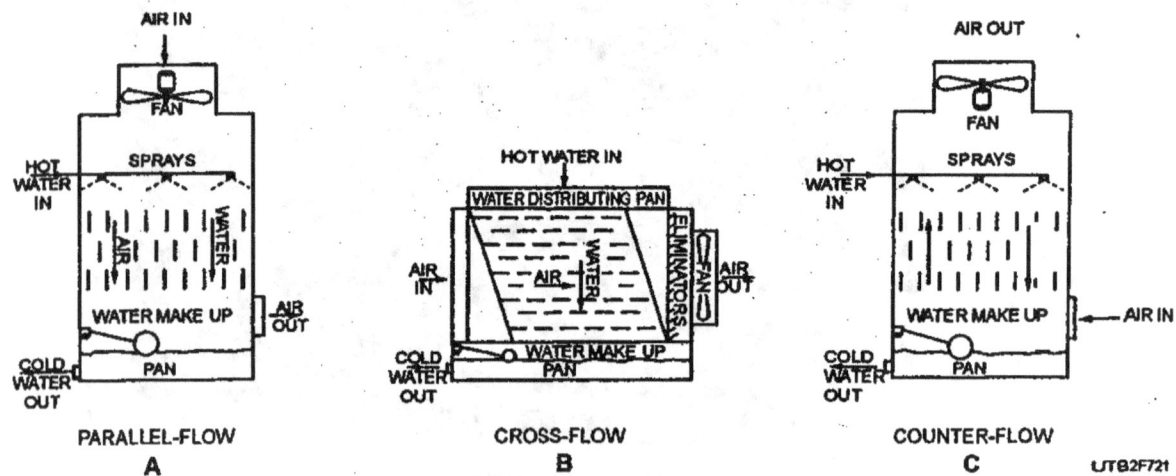

Figure 7-21.—Types of induced fan cooling towers.

with piping and wiring in place, it is simple to install and may be placed wherever there is a clearance of 2 feet for the intake end and a space of 10 feet or more in front of the fan. The discharge end must not face the prevailing wind and should not be directed into a traffic area because drift loss may be objectionable.

In some situations, an indoor location for the cooling tower may be desirable. An induced draft tower of the counterflow or cross-flow design is generally selected for indoor installation. Two connections to the outside are usually required—one for drawing outdoor air into the tower and the other for discharging it back to the outside. A centrifugal blower is often necessary for this application to overcome the static pressure of the ductwork. Many options are possible as to the point of air entrance and air discharge. This flexibility is often important in designing an indoor installation. Primary water breakup is by pressure spray and fill of various types.

The induced draft cooling tower for indoor installation is a completely assembled packaged unit but is so designed that it can be partially disassembled to permit passage through limited entrances. Indoor installations of cooling towers are becoming more popular. External space restrictions, architectural compatibility, convenience for observation and maintenance all combine to favor an indoor location. The installation cost is somewhat higher than an outdoor location. Packaged towers are available in capacities to serve the cooling requirements of refrigeration plants in the 5- to 75-ton range.

### Forced Draft

A forced draft cooling tower uses a fan to force air into the tower. In the usual installation, the fan shaft is in a horizontal plane. The air is forced horizontally through the fill and upward to be discharged out of the top of the tower.

Underflow cooling towers are an improved design of the forced draft tower that retains all the advantages of the efficient parallel-flow design. Air is forced into the center of the tower at the bottom. The air is then turned horizontally (both right and left) through fill chambers and is discharged vertically at both ends. By forcing the air to flow upward and outward through the fill and leave at the ends, operating noise is baffled and a desirable reduction of sound level is achieved. All sides of the underflow tower are smoothly encased with no louver openings. This blends with modern architecture and eliminates the necessity of masonry walls or other screening devices oftentimes necessary to conceal cooling towers of other types.

### Materials

Redwood has been the standard construction material for cooling towers for many years. Though cypress, as well as treated fir and pine, has been used occasionally, these materials have not enjoyed a wide application. Casings are constructed of laminated waterproof plywood. Such casings, as well as other noncorrosive materials at critical points, are essential in areas having a highly corrosive atmosphere. Nails, bolts, and nuts of copper or aluminum are almost standard practice for cooling tower construction.

Cooling towers of metal coated with plastic or bituminous materials that have air intake louvers and fill made of redwood have met with only limited success. The limited success is primarily because of the high maintenance cost as compared to wood towers.

Packaged towers with metal sides and wood fill are reasonably common. Some manufacturers have used sheet aluminum for siding for limited periods of time. Plastic slats have been used for fill material but have not proved satisfactory in all cases.

Fire ordinances of a large city may require that no wood be used in construction of cooling towers. With steel or some other fireproof casing and without fill, a cooling tower will comply with the most restrictive ordinances.

### Maintenance

Recently, cooling towers have been linked to the spread of Legionnaire's disease. Several precautionary measures are recommended to help eliminate this problem. These include placing of cooling towers downwind and use of chloride compounds as disinfectants on a monthly maintenance schedule.

Water treatment is an important part of the operation of a cooling tower. The evaporation of water from a cooling tower leaves some solids behind. Recirculation of the water in the condenser cooling tower circuit, and the accompanying evaporation, causes the concentration of solids to increase. This concentration must be controlled or scale and corrosion will result.

Though draining the system from time to time and refilling with fresh water is one method of control, it is not recommended. Soon after refilling, the dissolved solids again build up to a dangerous concentration. A more common practice is to waste a certain amount of water continually from the system to the sewer. The water wasted is called blowdown. Blowdown is sometimes accomplished by wasting sump water through an overflow. A better practice, however, is to bleed the required quantity of blowdown from the warm water leaving the condenser on its way to the cooling tower. A mineral salt buildup (calcium bicarbonate concentration) of 10 grains per gallon is considered the maximum allowable concentration for untreated water in the sump if serious corrosion and scaling difficulties are to be avoided.

Cooling towers evaporate about 2 gallons of water every hour for each ton of refrigeration. A gallon of water weighs 8.3 pounds, and about 1,000 Btu is needed to evaporate 1 pound of water. Thus, to evaporate a gallon of water, 8.3 x 1,000 or 8,300 Btu is required.

In many instances, the makeup water contains dissolved salts in excess of 10 grains per gallon. It is obvious, then, that even 100 percent blowdown will not maintain a sump concentration of 10 grains. If the blowdown alone cannot maintain satisfactory control, then chemicals should be used.

Makeup water for a cooling tower is the sum of drift loss, evaporation, and blowdown. The drift loss for mechanical draft towers ranges from 0.1 percent of the total water being cooled for the better designed towers to as much as 0.3 percent. In estimating makeup water for a cooling tower, the higher value of 0.3 percent for drift loss is suggested. If the drift loss is actually less than this, the excess makeup water supplied is merely wasted down the overflow. This does, in effect, increase the amount of blowdown and is favorable from the viewpoint that the concentration of scale-forming compounds in the tower sump will be somewhat lower.

Redwood is a highly durable material; however, it is not immune to deterioration. The type of deterioration varies with the nature of the environmental conditions to which the wood is exposed. The principal types of deterioration are leaching, delignification, and microbiological attack.

Algae and slime are present in water and must be controlled chemically or the rate of heat transfer in the condenser will be materially reduced. Condenser tubing, cooling tower piping, and metal surfaces in the water-circulating system must be protected from scale and corrosion.

Using too much of a chemical or using the wrong chemical is known as overtreatment. It can materially reduce the performance or the life of a cooling tower condenser circuit.

## COMPRESSORS

A compressor is the machine used to withdraw the heat-laden refrigerant vapor from the evaporator, compress it from the evaporator pressure to the condensing pressure, and push it to the condenser. A compressor is merely a simple pump that compresses the refrigerant gas. Compressors may be divided into the following three types—reciprocating, rotary, and centrifugal. The function of compressing a refrigerant is the same in all three general types, but the mechanical means differ considerably. Rotary compressors are used in small sizes only, and their use is limited almost exclusively to domestic refrigerators and small water coolers. Centrifugal compressors are used in large refrigerating and air-conditioning systems (fig. 7-22).

### Reciprocating Compressors

Reciprocating compressors are usually powered by electric motors, although gasoline, diesel, and turbine drivers are sometimes used. In terms of capacity, reciprocating compressors are made in fractional horsepower for small, self-contained air conditioners and refrigeration equipment, increasing in size to about 250 tons or more capacity in larger installations. Reciprocating compressors are furnished in open, semisealed, and sealed (hermetic) types.

OPEN.—An open type of compressor shaft is driven by an external motor. The shaft passes through the crankcase housing and is equipped with a shaft seal to prevent refrigerant and oil from leaking or moisture and air from entering the compressor. Pistons are actuated by crankshafts or eccentric drive mechanisms mounted on the shaft. Discharge valves are usually mounted in a plate over the pistons. Suction valves are usually mounted either in the pistons, if suction vapors enter the cylinder through the side of the cylinder or through the crankcase, or in the valve plate over the pistons, if suction vapors enter the cylinder through the head and valve plate.

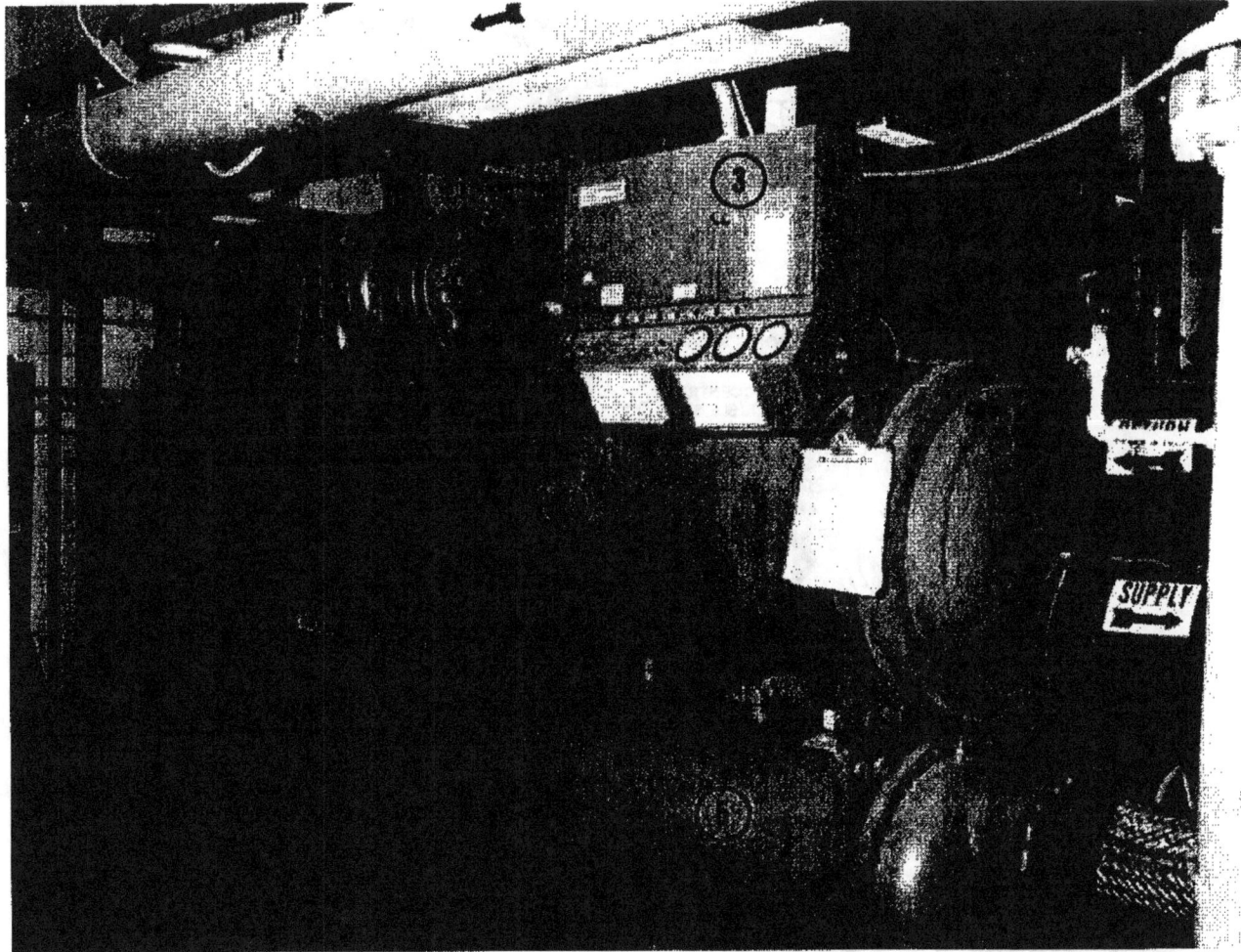

1. Centrifugal compressor with 5-inch impeller.
2. Refrigerant colled motor.
3. Control center.
4. Chiller section.
5. Condenser section.

Figure 7-22.—High-speed (36,000 rpm) single-stage centrifugal chiller.

Figure 7-23 shows a cross section of a typical open type of eccentric shaft compressor with suction valves in the valve plate of the head. Most belt-driven open type of compressors under 3 horsepower use a splash feed lubrication, but in larger size compressors, forced feed systems having positive displacement oil pumps are more common. The oil pump is usually driven from the rear end of the main shaft. Oil from the crankcase is forced under pressure through a hole in the main shaft to the seal, main bearing, and rod bearing, and through a hole in the rod up to the piston pins. Hermetically sealed compressor units used in window air conditioners are quite common in commercial sizes (under 5 horsepower) and are even made by some manufacturers in large tonnage sizes.

**SEMISEALED.**—Semisealed compressors are sometimes made in small sizes, but large tonnage units are always of the semisealed type. The primary difference between a fully sealed and a semisealed motor compressor is that in semisealed types the valve plates, and in some units the oil pump, can be removed for repair or replacement. This type of construction is helpful in larger sizes that are so bulky they would cause considerable trouble and expense in shipping, removing, and replacing the unit as a whole. Figure 7-24 shows a small semisealed compressor.

Sealed or semisealed units eliminate the belt drive and crankshaft seal, both of which are among the chief causes of service calls. Sealed and semisealed compressors are made either vertical or horizontal. The vertical type (fig. 7-25) usually has a positive displacement oil pump that forces oil under pressure of 10 to 30 psi to the main bearings, rod, or eccentric and pins, although they are sometimes splash oiled.

Although oil pumps for forced feed lubrication are also used on horizontal hermetic compressors, oil

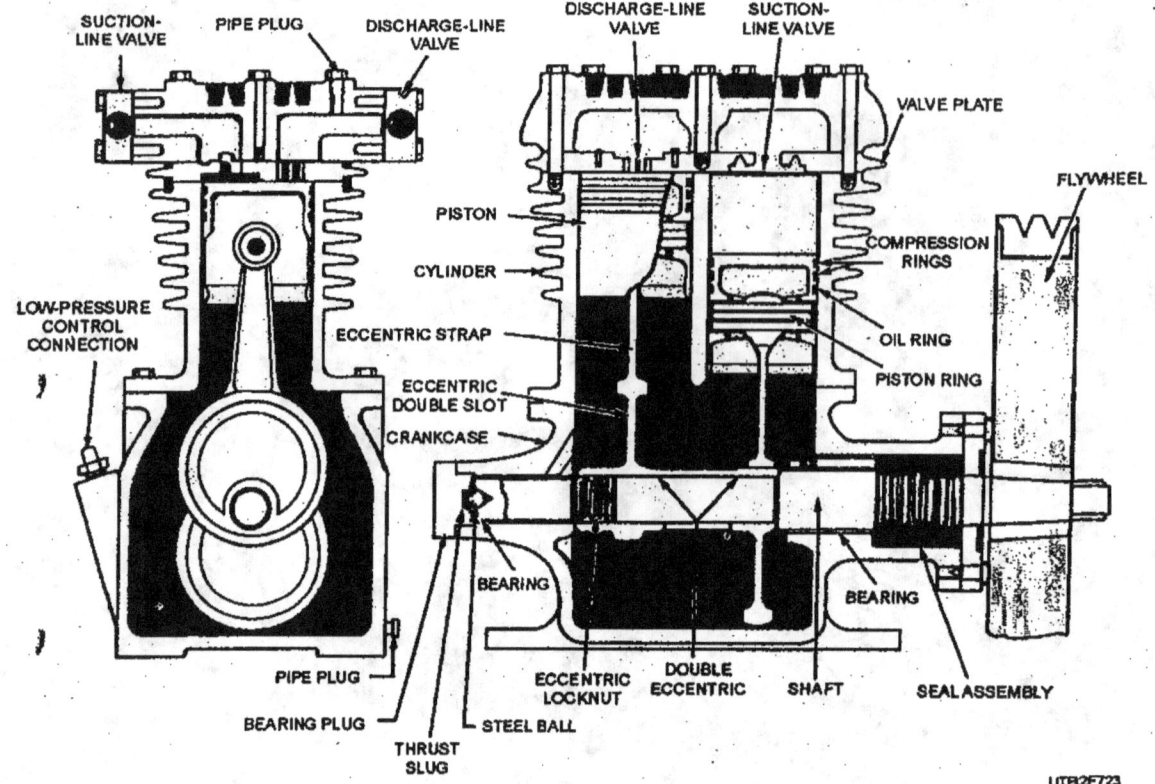

Figure 7-23.—Cross section of an open type of reciprocating compressor.

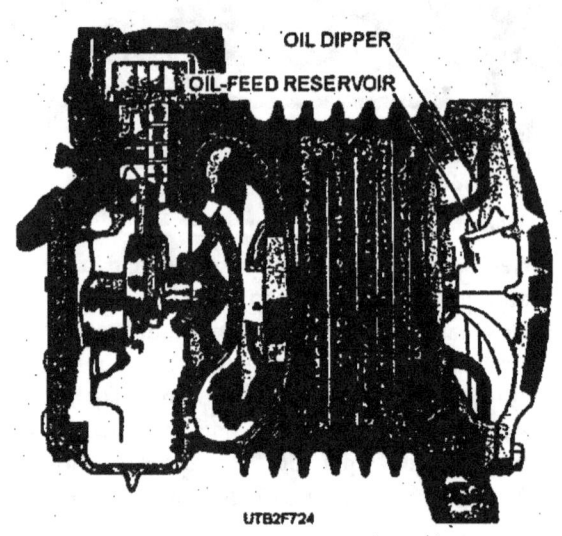

Figure 7-24.—Small semisealed compressor.

circulation at low oil pressure may be provided by slingers, screw type of devices, and the like. Splash and other types of oil feed must hot be considered inferior forced feed. With good design, they lubricate well. It is most important to maintain the proper oil level, use a correct grade of oil, and keep the system clean and free of dirt and moisture. This is true for all compression refrigeration systems, especially those equipped with

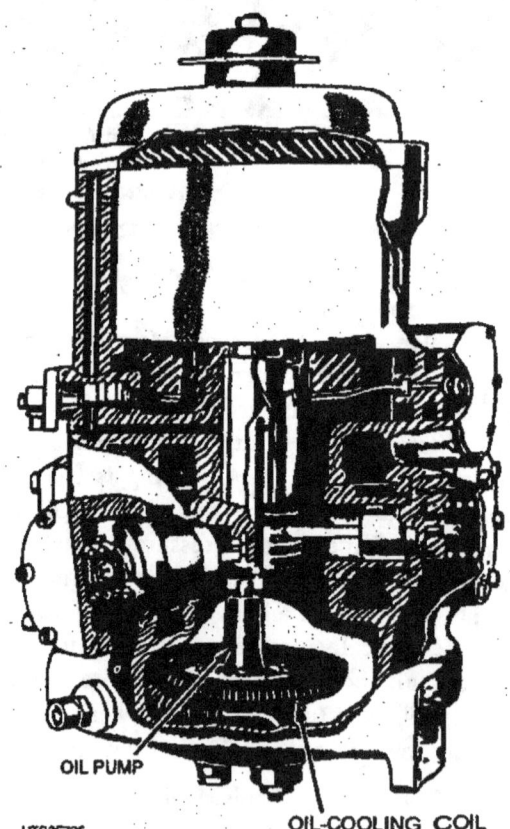

Figure 7-25.—Vertical semisealed compressor.

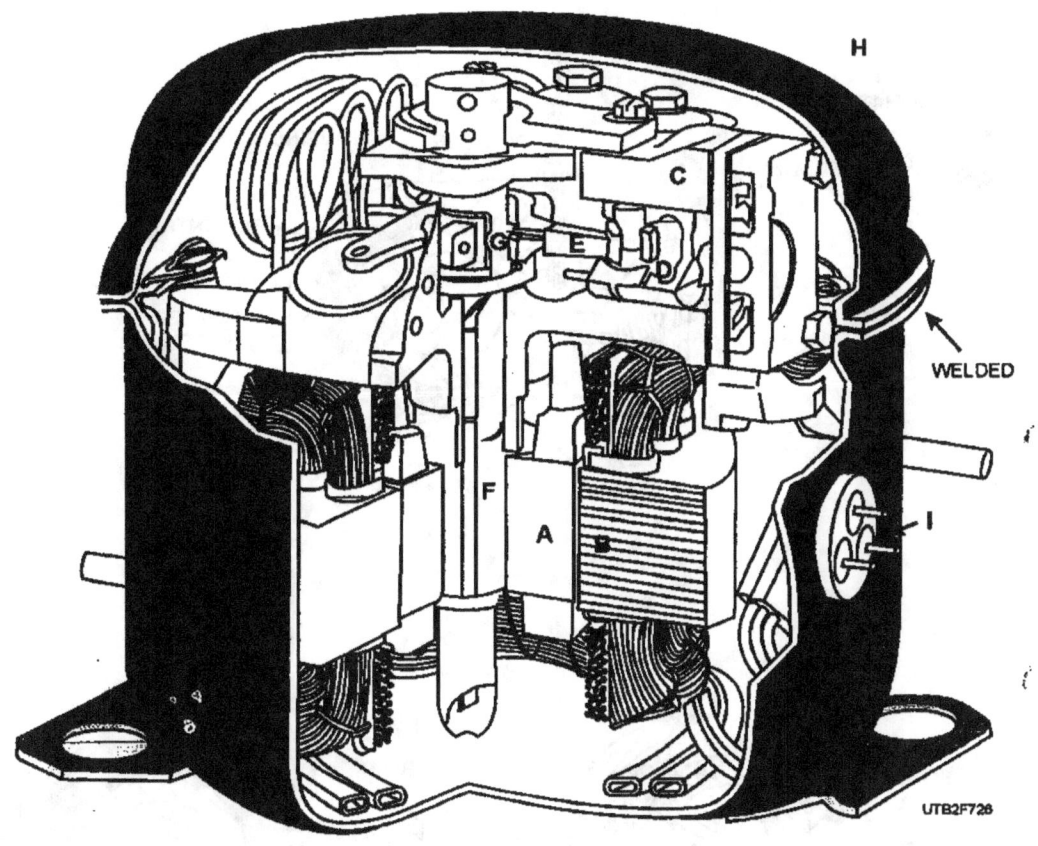

Figure 7-26.—Reciprocating hermetic compressor. (A) Motor rotor; (B) Motor stator; (C) Compressor cylinder; (D) Compressor piston; (E) Connecting rod; (F) crankshaft; (G) Crank throw; (H) Compressor shell (I) Glass sealed electrical connection.

hermetically sealed units whose motor windings may be attacked by acids or other corrosive substances introduced into the system or formed by the chemical reaction of moisture, air, or other foreign substances.

**HERMETIC.**—The term *sealed* or *hermetic* unit merely means that the motor rotor and compressor crankshaft of the refrigeration system are made in one piece, and the entire motor and compressor assembly is put into a gastight housing that is welded shut (fig. 7-26). This method of assembly eliminates the need for certain parts found in the open unit. These parts are as follows: motor pulley, belt, compressor flywheel, and compressor seal. The elimination of the preceding parts in the sealed unit similarly does away with the following service operations: replacing motor pulleys, replacing flywheels, replacing belts, aligning belts, and repairing or replacing seals. When it is realized there are major and minor operations that maintenance personnel must perform and the sealed unit dispenses with only five of these, it can be readily seen that servicing is still necessary.

**Rotary Compressors**

Rotary compressors are generally associated with refrigerators, water coolers, and similar small capacity equipment. However, they are available in larger sizes. A typical application of a large compressor is found in compound compressor systems where high capacity must be provided with a minimum of floor space.

In a rotary compressor (fig. 7-27), an eccentric rotor revolves within a housing in which the suction and discharge passages are separated by means of a sealing blade. When the rotating eccentric first passes this blade, the suction area is at a minimum. Further rotation enlarges the space and draws in the charge of refrigerant. As the eccentric again passes the blade, the gas charge is shut off at the inlet, compressed, and discharged from the compressor. There are variations

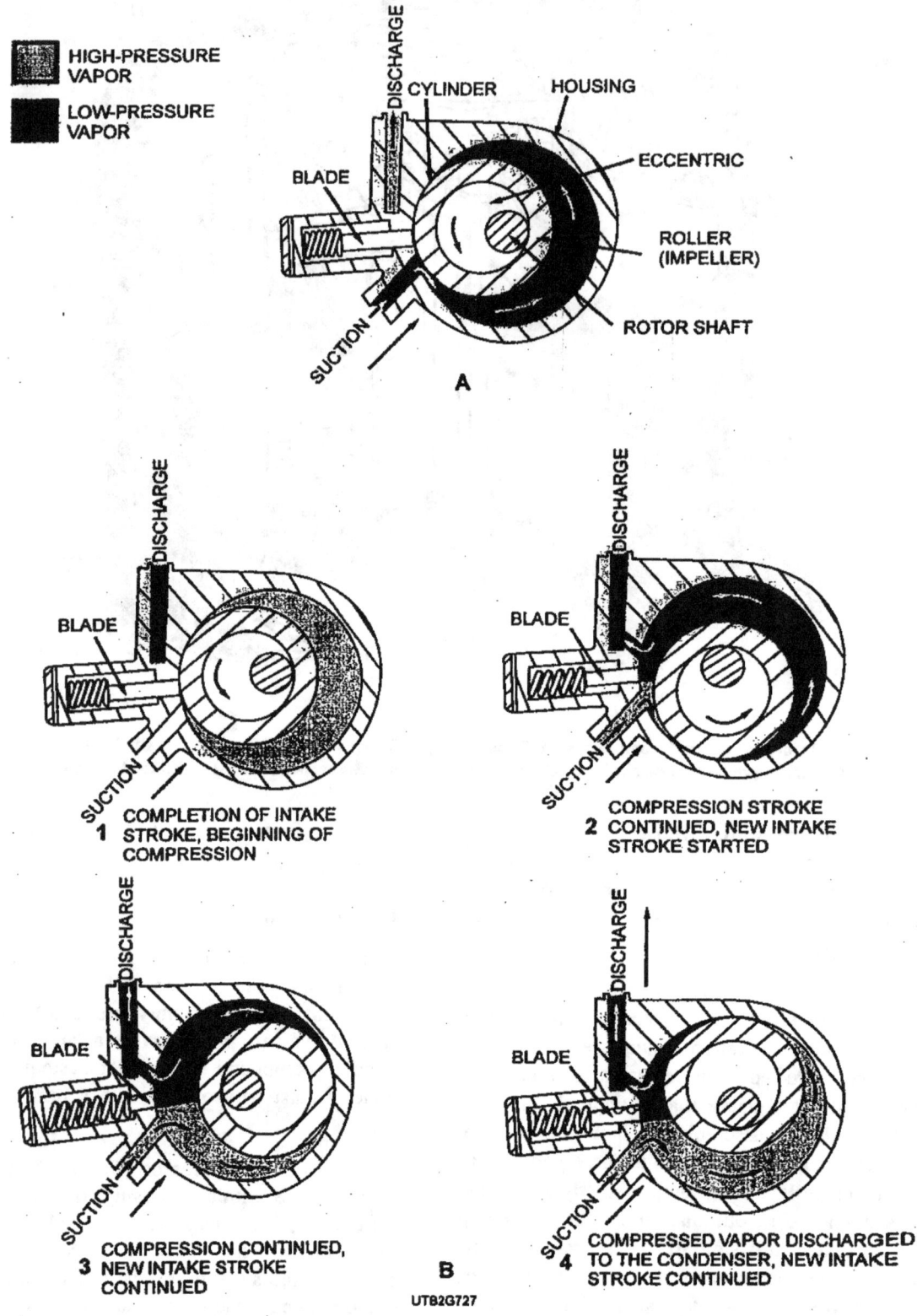

Figure 7-27.—Rotary compressor: A. Part identification; B. Operation.

of this basic design, some of which provide the rotor with blades to trap and compress the vapor.

## Centrifugal Compressors

Centrifugal compressors are used in large refrigeration and air-conditioning systems, handling large volumes of refrigerants at low-pressure differentials. Their operating principles are based on the use of centrifugal force as a means of compressing and discharging the vaporized refrigerant. Figure 7-28 is a cutaway view of one type of centrifugal compressor. In this application, one or two compression stages are used, and the condenser and evaporator are integral parts of the unit. The heart of this type of compressor is the impeller wheel.

## Scroll Compressors

A scroll compressor has two different offset spiral disks to compress the refrigerant vapor. The upper scroll is stationary, while the lower scroll is the driven scroll. Intake of refrigerant is at the outer edge of the driven scroll, and the discharge of the refrigerant is at the center of the stationary scroll. The driven scroll is rotated around the stationary or "fixed" scroll in an orbiting motion. During this movement, the refrigerant vapor is trapped between the two scrolls. As the driven scroll rotates, it compresses the refrigerant vapor through the discharge port. Scroll compressors have few moving parts and have a very smooth and quiet operation.

## CONTROLS

Controls used in air conditioning are generally the same as for refrigeration systems—thermostats, humidistats, pressure and flow controllers, and motor overload protectors (fig. 7-29).

### Thermostats

The thermostat is an adjustable temperature-sensitive device, which through the opening and closing of its contacts controls the operation of the

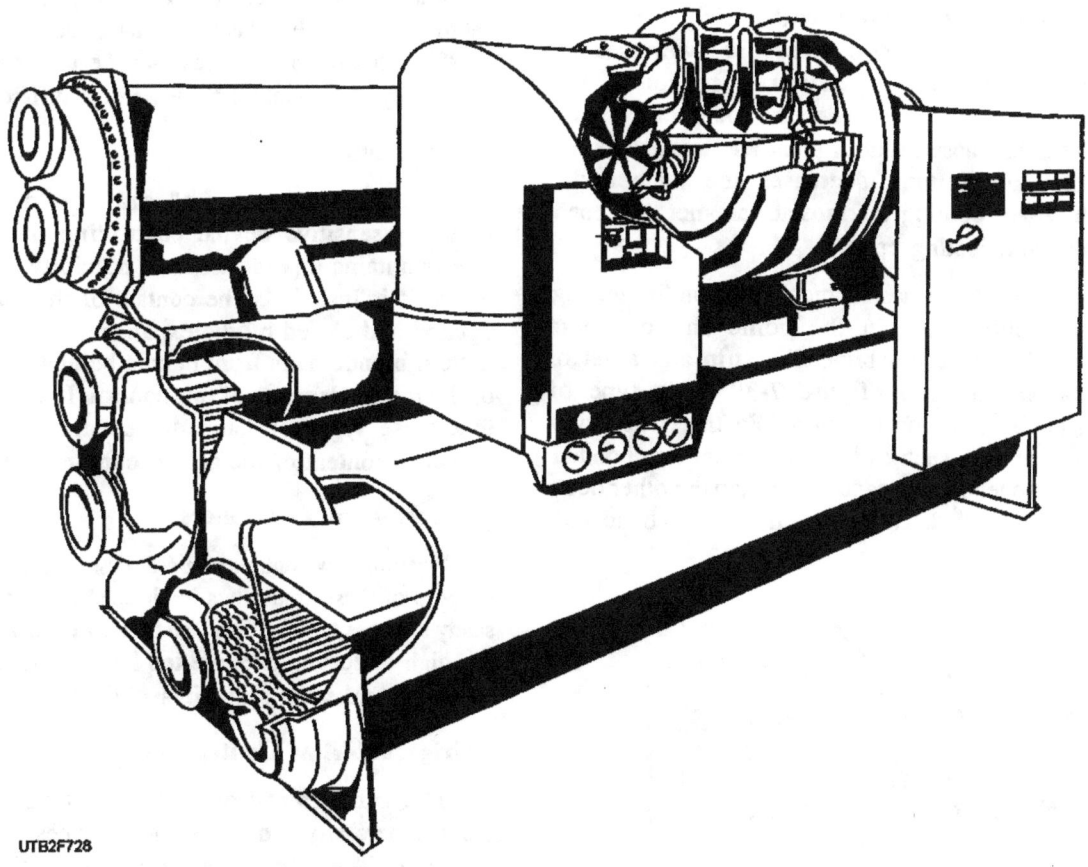

Figure 7-28.—Cutaway view of one type of centrifugal compressor.

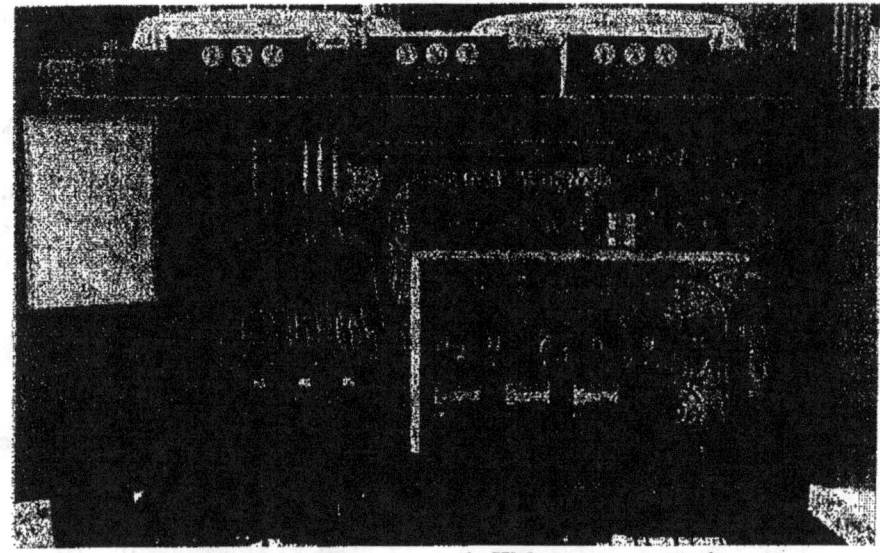

1. Commpressor breakers.
2. Compressor starters.
3. Fan cycle controls.
4. High-pressure controls.
5. Oil failure controls.
6. Solid-state staging thermostat.

Figure 7-29.—Packaged air-cooled chiller controls.

cooling unit. The temperature-sensitive element may be a bimetallic strip or a confined, vaporized liquid.

The thermostats used with refrigerative air conditioners are similar to those used with heating equipment, except their action is reversed. The operating circuit is closed when the room temperature rises to the thermostat control point and remains closed until the cooling unit decreases the temperature enough. Also, cooling thermostats are not equipped with heat-anticipating coils.

Wall type of thermostats most common for heating and air conditioning in the home and on some commercial units use a bimetallic strip and a set of contacts, as shown in figure 7-30. This type of thermostat operates on the principle that when two dissimilar metals, such as brass and steel, are bonded together, one tends to expand faster than the other does when heat is applied. This causes the strip to bend and close the controls.

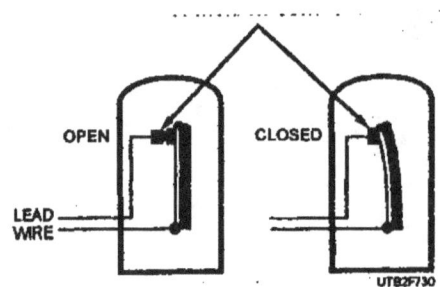

Figure 7-30.—Bimetallic thermostat.

As a Utilitiesman, you may be required to make an adjustment that sets the temperature difference between the cut-in and cutout temperatures. For example, if the system is set to cut in at 76°F and cut out at 84°F, then the differential is 8°F. This condition prevents the unit from cycling continually as it would if there were no differential.

### Humidistats

A room "humidistat" may be defined as a humidity-sensitive device controlling the equipment that maintains a predetermined humidity of the space where it is installed. The contact of the humidistat is opened and closed by the expansion or contraction of natural blonde hairs from human beings, which is one of the major elements of this control. It has been found that these types of hairs are most sensitive to the moisture content of the air surrounding them.

### Pressure-Flow Controllers

Pressure-flow controllers are discussed in chapter 6. The purpose of these controllers in air conditioning is to act as safety switches for the system, so if either the head pressure is too high or suction pressure too low, the system will be secured regardless of the position of the operating switches.

### Refrigerant-Flow Controllers

The refrigerant-flow controllers used with air conditioners are also similar to the ones discussed in chapter 6. These controllers are either of the capillary type or externally equalized expansion valve type and are usually of larger tonnage than those used for refrigerators.

## Motor Overload Protectors

When the compressor is powered by an electric motor, either belt driven or as an integral part of the compressor assembly, the motor is usually protected by a heat-actuated overload device. This is in addition to the line power fuses. The heat to actuate the overload device is supplied by the electrical energy to the motor, as well as the heat generated by the motor itself. Either source of heat or a combination of the two, if too much, causes the overload device to open and remove the motor from the line.

Figure 7-31 shows a thermal-element type of overload cutout relay. It is housed in the magnetic starter box. On current overload, the relay contacts open, allowing the holding coil to release the starting mechanism, thereby stopping the motor.

An oil failure cutout switch is provided on many systems to protect the compressor against oil failure. The switch is connected to register pressure differential between the oil pump and the suction line. Figure 7-32 shows a typical oil failure cutout switch. The switch contains two bellows, which work against each other, and springs for adjusting. Tubing from the oil pump is connected to the bottom bellows of the switch. Tubing from the suction line is connected to the upper bellows. When a predetermined pressure differential is not maintained, a pair of contacts in the switch is opened and breaks the circuit to the compressor motor. A heating element with a built-in delay is in the switch to provide for starting the compressor when oil pressure is low.

The water-regulating valve used with a water-cooled condenser responds to a predetermined condensing pressure. A connection from the discharge side of the compressor to the valve transmits condensing pressure directly to a bellows inside the

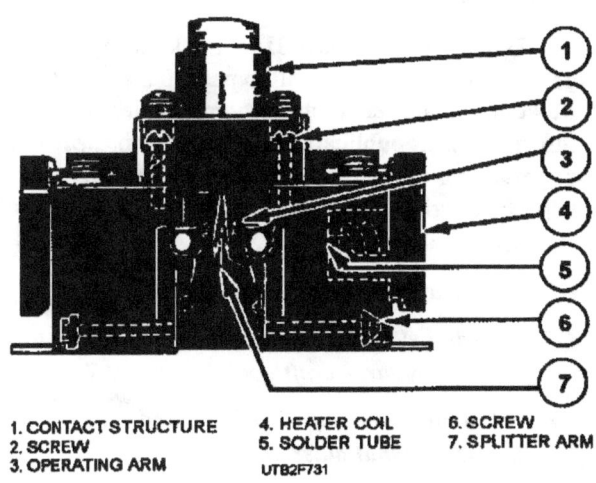

1. CONTACT STRUCTURE
2. SCREW
3. OPERATING ARM
4. HEATER COIL
5. SOLDER TUBE
6. SCREW
7. SPLITTER ARM

Figure 7-31.—Thermal overload relay.

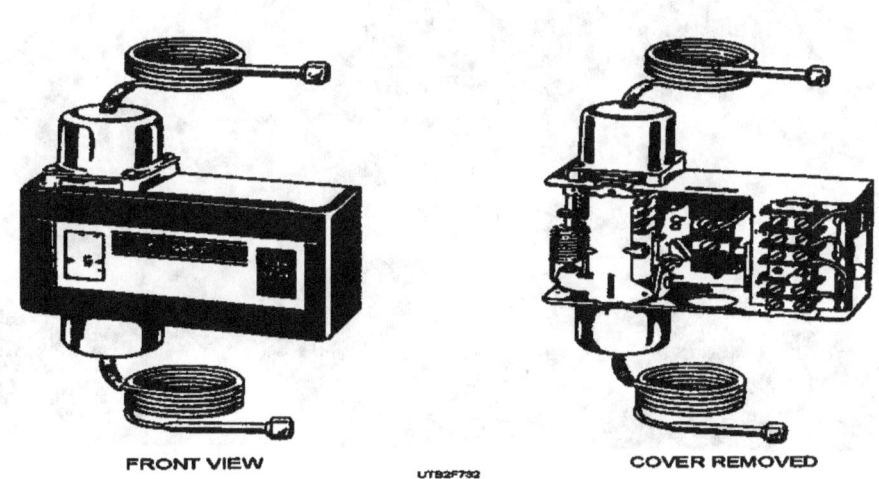

FRONT VIEW　　COVER REMOVED

Figure 7-32.—Oil failure cutout switch.

7-31

valve. High pressure opens the valve, allowing a greater flow of water; low pressure throttles the flow. Use of such a valve provides for a more economical use of water for condensing. Figure 7-33 shows a typical water-regulating valve. When condenser water is supplied by a cooling tower, water-regulating valves are not customarily used because the cooling tower fan and circulating pump are wired into the compressor motor control circuit.

### Step Controller

The step controller contains a shaft upon which is mounted a series of cams. Rotation of the cams, in turn, operates electrical switches. Through adjustment of the cams on the shaft, the temperature at which each switch is to close and open (differential) is established. In addition, the switches may be adjusted to operate in almost any sequence (fig. 7-34).

## TROUBLESHOOTING

Table Z of appendix II is a troubleshooting chart generally applicable to all types of air conditioners. Most manufacturers include more detailed and specific information in publications pertaining to their units. If you find that there is no manual with the unit when it is unpacked, write to the manufacturer and request one as soon as possible.

Q21. How are cooling towers classified?

Q22. A wind velocity of 8 mph is generally used to design natural draft cooling towers. True /False

Q23. Counter flow, parallel flow, and cross flow are types of what class of cooling tower?

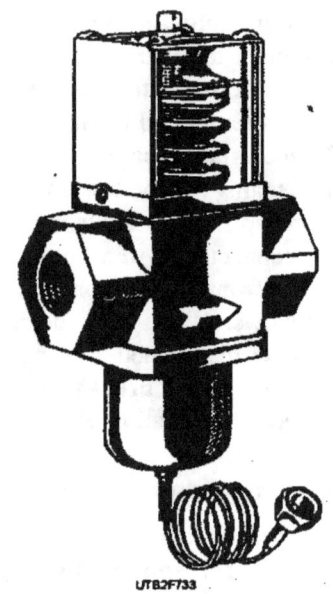

Figure 7-33.—Water-regulating valve.

Figure 7-34.—Step controller and pressure-sensor configuration. (A) Step controller with modulating motor, single-pole double-throw mini switches, and mouse trap relay assembly; (B) Pressure sensor that controls the step controller.

Q24. What type of cooling tower is installed indoors?

Q25. Forced draft underflow towers retain the advantages of what other type of cooling tower?

Q26. Air intake louvers and fill are made of what material?

Q27. Cooling towers evaporate approximately how much water every hour for each ton of refrigeration?

Q28. Rotary compressors are used in what type of units?

Q29. Semisealed and sealed compressors have reduced service requirements because of the elimination of what part?

Q30. What control is temperature sensitive and controls the operation of the cooling unit?

Q31. What device maintains humidity at a predetermined point?

Q32. What causes a motor to shut down when a motor is too hot?

# AUTOMOTIVE AIR CONDITIONING

**Learning Objective:** Understand the basic principles of operation, maintenance, and repair of automotive air conditioning.

Vehicle air conditioning is the cooling (refrigeration) of air within a passenger compartment. Refrigeration is accomplished by making practical use of three laws of nature—heat transfer, latent heat of vaporization, and the effects of pressure on boiling or condensation. The first two laws are discussed in chapter 6 of this TRAMAN; the practical application of the third is outlined below.

## EFFECT OF PRESSURE ON BOILING OR CONDENSATION

The saturation temperature (the temperature where boiling or condensation occurs) of a liquid or vapor increases or decreases according to the pressure exerted on it.

In the fixed orifice tube refrigerant system, liquid refrigerant is stored in the condenser under high pressure (fig. 7-35). When the liquid refrigerant is released into the evaporator by the fixed orifice tube,

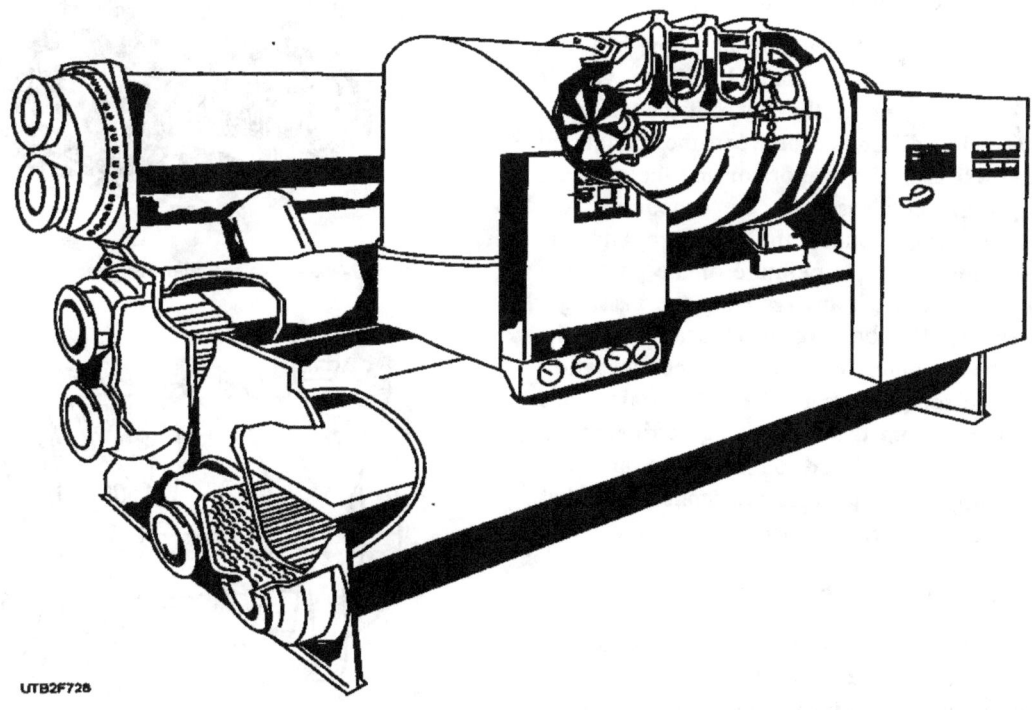

Figure 7-35.—Air-conditioning refrigeration system-fixed orifice.

7-33

the resulting decrease in pressure and partial boiling lowers its temperature to its new boiling point. As the refrigerant flows through the evaporator, passenger compartment air passes over the outside surface of the evaporator coils. As it boils, the refrigerant absorbs heat from the air and thus cools the passenger compartment. The heat from the passenger compartment is absorbed by the boiling refrigerant and hidden in the vapor. The refrigeration cycle is now under way. The following functions must be done to complete the refrigeration cycle:

1. Disposing of the heat in the vapor

2. Converting the vapor back to liquid for reuse

3. Returning of the liquid to the starting point in the refrigeration cycle

The compressor and condenser (fig. 7-35) perform these functions. The compressor pumps the refrigerant vapor (containing the hidden heat) out of the evaporator and suction accumulator drier, then forces it under high pressure into the condenser which is located in the outside air stream at the front of the vehicle. The increased pressure in the condenser raises the refrigerant condensation or saturation temperature to a point higher than that of the outside air. As the heat transfers from the hot vapor to the cooler air, the refrigerant condenses back to a liquid. The liquid under high pressure now returns through the liquid line to the fixed orifice tube for reuse.

It may seem difficult to understand how heat can be transferred from a comparatively cooler vehicle passenger compartment to the hot outside air. The answer lies in the difference between the refrigerant pressure that exists in the evaporator and the pressure that exists in the condenser. In the evaporator, the compressor suction reduces the pressure and the boiling point below the temperature of the passenger compartment. Thus heat transfers from the passenger compartment to the boiling refrigerant. In the condenser, the compressor raises the condensation point above the temperature of the outside air. Thus the heat transfers from the condensing refrigerant to the outside air. The fixed orifice tube and the compressor simply create pressure conditions that permit the laws of nature to function.

## AUTOMOTIVE COMPRESSORS

There are three basic types of air-conditioning compressors in general use in automotive applications. Each of these uses a reciprocating (back-and-forth motion) piston arrangement—two-cylinder reciprocating, swash plate, and scotch yoke. Most automotive compressors are semihermetic.

Two-cylinder compressors (fig. 7-36) usually contain two pistons in a parallel V-type configuration. The pistons are attached to a connecting rod, which is driven by the crankshaft. The crankshaft is connected to the compressor clutch assembly, which is driven by an engine belt. Reed valves generally are used to control the intake and exhaust of the refrigerant gas during the pumping operation. These compressors are usually constructed of die cast aluminum.

In the swash plate or "wobble plate" compressor (fig. 7-37), the piston motion is parallel to the

Figure 7-36.—Two-cylinder reciprocating compressor.

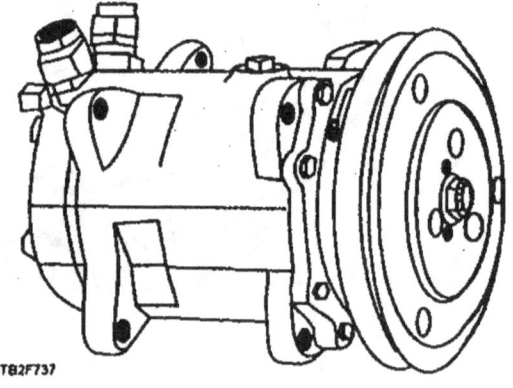

Figure 7-37.—Five-cylinder swash plate compressor.

crankshaft. The pistons are connected to an angled swash plate using ball joints. Swash plate compressors are of three types—five-cylinder, six-cylinder, and five-cylinder variable.

The five- and six-cylinder swash compressor has, in effect, three cylinders at each end of its inner assembly. A swash plate of diagonal design is mounted on the compressor shaft. It actuates the pistons, forcing them to move back and forth in the cylinders as the shaft is rotated. Reed valves control suction and discharge; crossover passages feed refrigerant to both high- and low-service fittings at the rear end of the compressor. A gear type of oil pump in the rear head provides for compressor lubrication.

The five-cylinder variable swash plate compressor is different from the other swash plate compressors. It uses a plate connected to a hinge pin that permits the swash plate to change its angle. The angle of the swash plate is controlled by a bellows valve that senses suction pressure. During high load conditions the swash plate angle is large, and during low load conditions, the swash plate is smaller. The displacement of the compressor is high at a large angle and low at a small angle.

A scotch-yoke compressor changes rotary motion into reciprocating motion. The basic mechanism of the scotch yoke contains four pistons mounted 90 degrees from each other. Opposed pistons are pressed into a yoke that rides on a slide block located on the shaft eccentric (fig. 7-38). Rotation of the shaft provides a reciprocating motion with no connecting rods. Refrigerant flows into the crankcase through the rear and is drained through the reeds attached to the piston tops during the suction stroke. Refrigerant is then discharged through the valve plate out the connector block at the rear. These compressors are shorter in length and larger in diameter than other compressors.

## Compressor Service Valves

Compressor service valves are built into some systems. They serve as a point of attachment for test gauges or servicing hoses. The service valves have three position controls—front seated, back seated, and midposition (fig. 7-39).

The position of this double-faced valve is controlled by rotating the valve stem with a service valve wrench. Clockwise rotation seats the front face of the valve and shuts off all refrigerant flow in the system. This position isolates the compressor from the rest of the system.

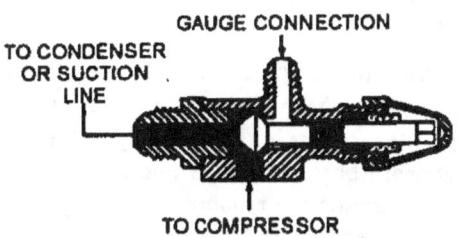
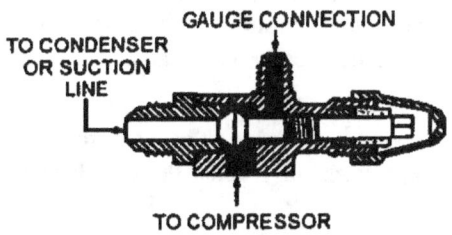
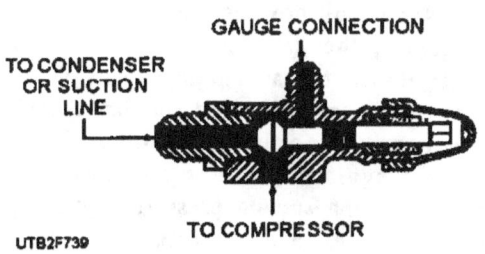

Figure 7-39.—Three-way service valve positions.

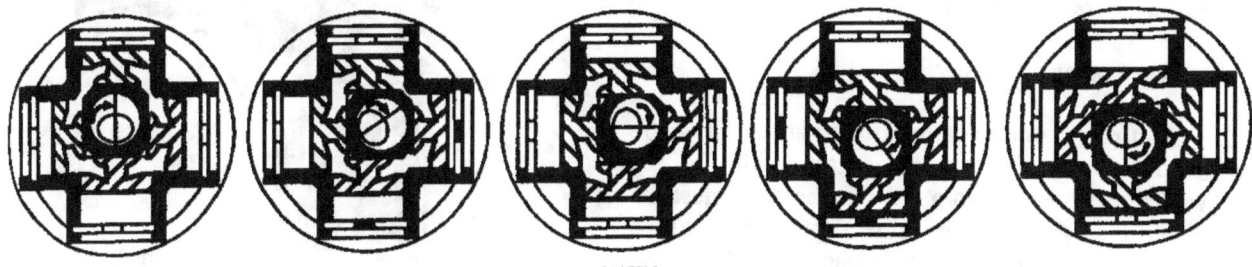

Figure 7-38.—Four-cylinder scotch-yoke mechanism.

7-35

Counterclockwise rotation unseats the valve and opens the system to refrigerant flow (midposition). Systematic checks are performed with a manifold gauge set with the service valve in midposition. Further counterclockwise rotation of the valve stem seats the rear face of the valve. This position opens the system to the flow of refrigerant but shuts off refrigerant to the test connector. The service valves are used for observing of operating pressures; isolating the compressor for repair or replacement; and discharging, evacuating, and charging the system.

Compressors used in automotive air-conditioning systems generally are equipped with an electromagnetic clutch that energizes and de-energizes to engage and disengage the compressor. Two types of clutches are in general use-the rotating coil and the stationary coil.

The rotating coil clutch has a magnetic coil mounted in the pulley that rotates with the pulley. It operates electrically through connections to a stationary brush assembly and rotating slip rings. The clutch permits the compressor to engage or disengage as required for adequate air conditioning. The stationary coil clutch has the magnetic coil mounted on the end of the compressor. Electrical connections are made directly to the coil leads.

The belt-driven pulley is always in rotation while the engine is running. The compressor is in rotation and operation only when the clutch engages it to the pulley.

Air-conditioning and refrigeration systems use various control devices, including those for the refrigerant, the capillary tube usually found on window units, the automatic expansion valves also found on window units and small package units, the thermal expansion valve, and various types of suction pressure-regulating valves and devices. A brief description of a suction pressure-regulating valve is given below. A suction pressure-regulating valve is used on automotive air conditioning because the varying rpm of the compressor unit must maintain a constant pressure in the evaporator.

### Suction Pressure-Regulating Valves

Suction pressure-regulating valves may be installed in the suction line at the outlet of the evaporator when a minimum temperature must be maintained. Suction pressure-regulating valves decrease the temperature difference, which would otherwise exist between the compartment temperature and the surface of the cooling coils. The amount of heat that can be transferred into the evaporating refrigerant is directly proportional to the temperature difference. Figure 7-40 shows an exploded view of a typical suction pressure-regulating valve, sometimes called a suction throttling valve in automotive air conditioners.

Three types of suction pressure-regulating valves are used—suction throttling valve (STV), evaporator pressure regulators (EPR), or pilot-operated absolute valve (POA), developed by General Motors Corporation. These valves, in most cases, are adjustable.

The POA valve uses a sealed pressure element that maintains a constant pressure independent of the altitude of the vehicle. There are two basic types of metering devices built into a single container—the VIR (Valves-In-Receiver) and the EEVIR (Evaporator Equalized Valves-In-Receiver). These units combine the POA valve, receiver-drier, thermostatic expansion valve, and sight glass into a single unit.

The VIR assembly is mounted next to the evaporator, which eliminates the need for an external equalizer line between the thermostatic expansion valve and the outlet of the POA valve. The equalizer function is carried out by a drilled hole (equalizer port) between the two-valve cavities in the VIR housing.

The thermostatic expansion valve is also eliminated. The diaphragm of the VIR expansion valve is exposed to the refrigerant vapor entering the VIR unit from the outlet of the evaporator. The sight glass is in the valve housing at the inlet end of the thermostatic valve cavity where it gives a liquid indication of the refrigerant level.

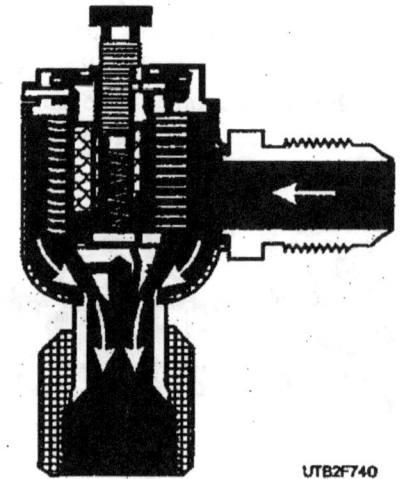

Figure 7-40.—A typical suction pressure-regulating valve.

The VIR thermostatic expansion valve controls the flow of refrigerant to the evaporator by sensing the temperature and pressure of the refrigerant gas, as it passes through the VIR unit on its way to the compressor. The POA valve controls the flow of refrigerant from the evaporator to maintain a constant evaporator pressure of 30 psi. The VIR and the POA valves are capsule type of valves. When found to be defective, you must replace the complete valve capsule.

The drier desiccant is in a bag in the receiver shell. It is replaceable by removing the shell and removing the old bag and installing a new bag of desiccant.

Service procedures for the VIR system differ in some respect from the service procedures performed on conventional automotive air-conditioning systems.

## SERVICE PRECAUTIONS

Observe the following precautions whenever you are tasked to service air-conditioning equipment:

- Never open or loosen a connection before discharging the system.

- A system that has been opened to replace a component or one which has discharged through leakage must be evacuated before charging.

- Immediately after disconnecting a component from the system, seal the open fittings with a cap or plug.

- Before disconnecting a component from the system, clean the outside of the fittings thoroughly.

- Do not remove the sealing caps from a replacement component until you are ready to install it.

- Refrigerant oil absorbs moisture from the atmosphere if it is left uncapped. Do not open an oil container until it is ready to use, and install the cap immediately after using. Store the oil only in a clean, moisture-free container.

- Before connecting to an open fitting, always install a new seal ring. Coat the fitting and seal with the refrigerant oil before connecting.

- When installing a refrigerant line, avoid sharp bends. Position the line away from the exhaust or any sharp edges that may chafe the line.

- Tighten the fittings only to specified torque. The copper and aluminum fittings that are used in refrigerant systems will not tolerate over-tightening.

- When disconnecting a fitting, use a wrench on both halves of the fitting to prevent twisting of refrigerant lines or tubes.

- Do not open a refrigerant system or uncap a replacement component unless it is as close as possible to room temperature. This prevents condensation from forming inside a component that is cooler than the surrounding air.

- Keep the service tools and work area clean. Contamination of a refrigerant system through careless work habits must be avoided.

## DIAGNOSIS, TESTING, AND SERVICING

Diagnosis is more than just following a series of interrelated steps to find the solution to a specific condition. It is a way of looking at systems that are not functioning the way they should and finding out why. Also, it is knowing how the system should work and whether it is working correctly. All good diagnosticians use the same basic procedures.

There are basic rules for diagnosis. If these rules are followed, the cause of the condition will usually be found the first time through the system.

1. Know the system; know how the parts go together. Also, know how the system operates and its limits, and what happens when something goes wrong. Sometimes this means comparing a system that is working properly with the one you are servicing.

2. Know the history of the system. How old or new is the system? What kind of treatment has it had? Has it been serviced in the past in such a manner that might relate to the present condition? What is the service history? A clue in any of these areas might save a lot of diagnosis time.

3. Know the probability of certain conditions developing. It is true that most conditions are caused by simple things, rather than by complex ones, and they occur in a fairly predictable pattern. Electrical problem conditions, for instance, usually occur at connections, rather than in components. An engine "no-start" is more likely to be caused by a loose wire or some component out of adjustment than a sheared off camshaft. Know the difference between impossible and improbable. Many good technicians have spent hours diagnosing a system because they thought certain failures were "impossible," only to find out the failures

eventually were just "improbable" and actually had happened. Remember, new parts are just that—new. It does not mean they are good functioning parts.

4. Don't cure the symptom and leave the cause. Recharging a refrigerant system may correct the condition of insufficient cooling, but it does not correct the original problem unless a cause is found. A properly working system does not lose refrigerant over time.

5. Be sure the cause is found; do not be fooled into thinking the cause of the problem has been found. Perform the proper tests; then double-check the results. The system should have been checked for refrigerant leaks. If no leaks were found, perform a leak test with the system under extremely high pressure. If the system performed properly when new, it had to have a leak to be low in charge.

6. No matter what form charts may take, they are simply a way of expressing the relationship between the basic logic and a physical system of components. It is a way of determining the cause of a condition in the shortest possible amount of time. Diagnosis charts combine many areas of diagnosis into one visual display that allows you to determine the following:

- The probability of certain things occurring in a system
- The speed of checking certain components, or functions, before others
- The simplicity of performing certain tests before others
- The elimination of checking huge sections of a system by performing simple tests
- The certainties of narrowing down the search to a small area before performing in-depth testing

The fastest way to find a condition is to work with the tools that are available, which means working with proven diagnosis charts and the proper special tools for the system being worked on.

Servicing procedures for automotive air-conditioning units are similar to those used to service conventional air-conditioning systems. Discharging, evacuating, charging procedures, connections, and positions of valves on the gauge manifold set are shown in figure 7-41.

Servicing procedures for the VIR system are also similar to those used when servicing conventional air-conditioning systems. However, the hookup of the manifold gauge set is to the VIR unit. The high-pressure fitting is located in the VIR inlet line. The low-pressure fitting is located in the VIR unit.

## SYSTEM VISUAL INSPECTION

It is often possible to detect a problem caused by a careful visual inspection of the air-conditioning refrigerant system. This includes broken belts, obstructed condenser air passages, a loose clutch, loose or broken mounting brackets, disconnected or broken wires, and refrigerant leaks.

A refrigerant leak usually appears as an oily residue at the leakage point in the system. The oily residue soon picks up dust or dirt particles from the surrounding air and appears greasy. Through time, this builds up and appears to be heavy, dirt-impregnated grease.

Most common leaks are caused by damaged or missing O-ring seals at various hose and component connections. When these O rings are replaced, the new O rings should be lubricated with refrigerant oil. Care should be taken to keep lint from shop towels or cloths from contaminating the internal surfaces of the connection. Leakage may occur at a spring lock coupling if the wrong O rings are used at the coupling.

Another type of leak may appear at the internal Schrader type of air-conditioning charging valve core in the service gauge port valve fittings. If tightening the valve core does not stop the leak, it should be replaced with a new air-conditioning charging valve core.

Missing service gauge port valve caps can also cause a refrigerant leak. If this important primary seal (the valve cap) is missing, dirt enters the area of the air-conditioning charging valve core. When the service hose is attached, the valve depressor in the end of the service hose forces the dirt into the valve seat area, and it destroys the sealing surface of the air-conditioning charging valve core. When a service gauge port valve cap is missing, the protected area of the air-conditioning charging valve core should be cleaned and a new service gauge port valve cap should be installed.

### CAUTION

The service gauge port valve cap must be installed finger tight. If tightened with pliers, the sealing surface of the service gauge port valve may be damaged.

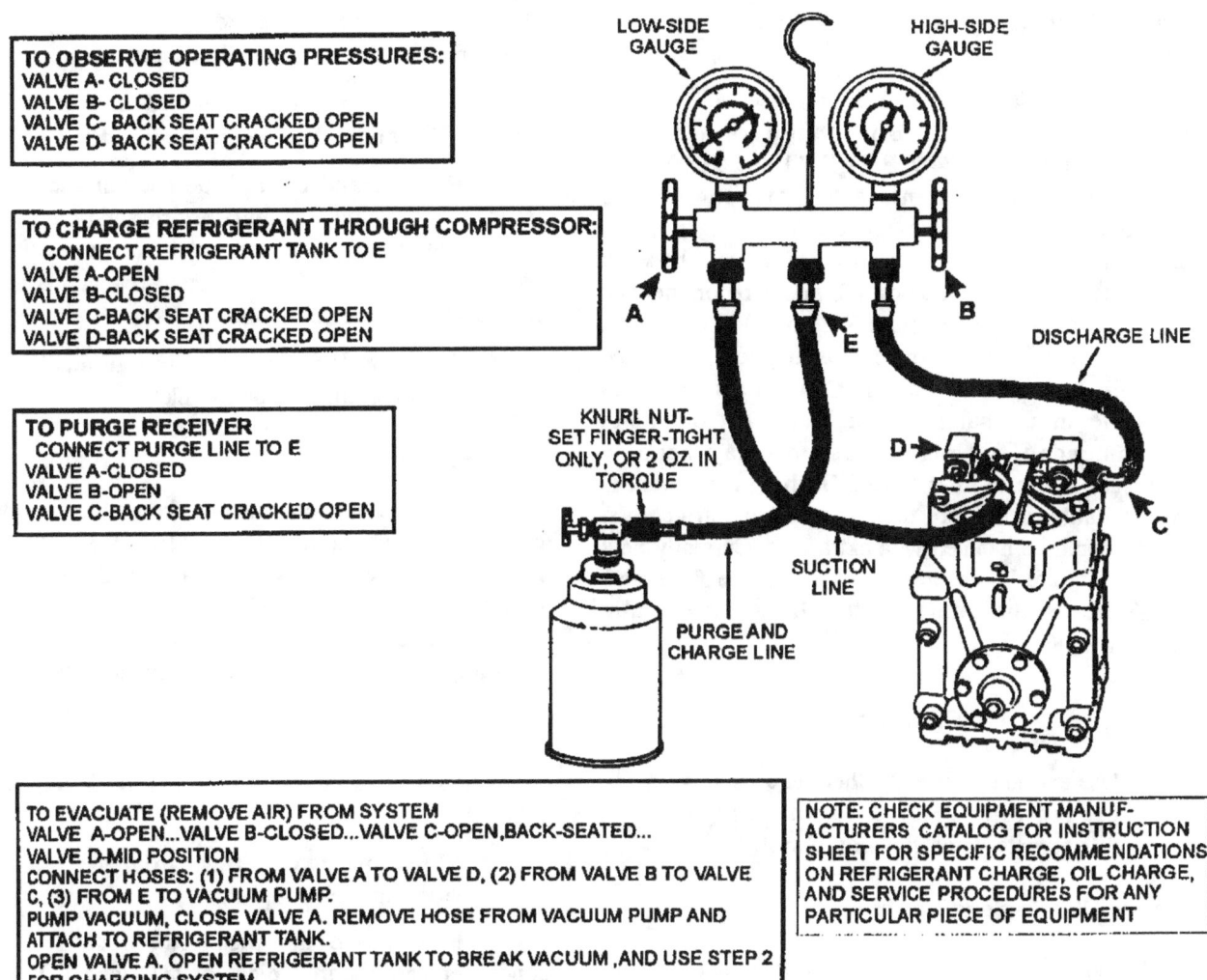

Figure 7-41.—Procedures for observing operating pressures, charging, purging, and evacuating a unit.

## CLEANING A BADLY CONTAMINATED REFRIGERANT SYSTEM

A refrigerant system can become badly contaminated for a number of reasons.

- The compressor may have failed due to damage or wear.
- The compressor may have been run for some time with a severe leak or an opening in the system.
- The system may have been damaged by a collision and left open for some time.
- The system may not have been cleaned properly after a previous failure.
- The system may have been operated for a time with water or moisture in it.

A badly contaminated system contains water, carbon, and other decomposition products. When such a condition exists, the system must be flushed with a special flushing agent, using equipment designed especially for this purpose. Follow the suggestions and procedures outlined for proper cleaning.

### Flushing Agents

A refrigerant to be suitable as a flushing agent must remain in the liquid state during the flushing operation to wash the inside surfaces of the system components. Refrigerant vapor will not remove contaminant particles. They must be flushed with a liquid. Some refrigerants are better suited for this purpose than others.

R-11 and R-113 are suited for use with special flushing equipment. Both have rather high vaporization points—74.7°F for R-11 and 117.6°F for R-113. Both refrigerants also have low closed container pressures. This reduces the danger of an accidental system discharge to a ruptured hose or fitting. R-113 will do the best job and is recommended as a flushing refrigerant. Both R-11 and R-113 require a propellant or a pump type of flushing equipment due to their low closed container pressures. R-11 is available in pressurized containers. Although not recommended for regular use, it may become necessary to use R-11 if special flushing equipment is not available. It is more toxic than other refrigerants, and it should be handled with extra care. Currently new refrigerants are being developed to replace R-11 and R-113 because these refrigerants will be phased out by the year 2000.

### CAUTION

Use extreme care and adhere to all safety precautions related to the use of refrigerants when flushing a system.

### System Cleaning and Flushing

When it is necessary to flush a refrigerant system, the suction accumulator/drier must be removed and replaced, as it is impossible to clean. Remove the fixed orifice tube. If a new tube is available, replace the contaminated one; otherwise, wash it carefully in flushing refrigerant or mineral spirits and blow it dry. If it does not show signs of damage or deterioration, it may be reused. Install new O rings.

Any moisture in the evaporator will be removed during leak testing and system evacuation following the cleaning job. Perform each step of the cleaning procedure carefully as outlined below.

1. Check the hose connections at the flushing cylinder outlet and flushing nozzle to ensure they are secure.

2. Ensure the flushing cylinder is filled with approximately 1 pint of R-113 and that the valve assembly on top of the cylinder is tightened securely.

3. Connect a can of R-12 or R-134a to the Schrader valve at the top of the charging cylinder. A refrigerant hose and a special, safety type of refrigerant dispensing valve are required for connecting the small can to the cylinder. Ensure all connections are secure.

4. Connect a gauge manifold and a discharge system. Disconnect the gauge manifold.

5. Remove and discard the suction accumulator/drier. Install a new accumulator/drier and connect it to the evaporator. Do not connect it to the suction line from the compressor. Ensure a protective cap is in place on the suction line connection.

6. Replace the fixed orifice tube. Install a protective cap on the evaporator inlet tube as soon as the new orifice tube is in place. The liquid line will be connected later.

7. Remove the compressor from the vehicle for cleaning and servicing or replacement, whichever is required. If the compressor is cleaned and serviced, add the specified amount of refrigerant oil before installing it on the mounting brackets in the vehicle. Install the shipping caps on the compressor connections. Install a new compressor on the mounting brackets in the vehicle.

8. Back flush the condenser and the liquid line as follows:

    a. Remove two O rings from the condenser inlet tube spring lock coupling.

    b. Remove the discharge hose from the condenser and clamp a piece of (1/2-inch ID) heater hose to the condenser inlet line. Ensure the hose is long enough to insert the free end into a suitable waste container to catch the flushing refrigerant.

    c. Move the flushing equipment into position and open the valve on the can of R-12 or R-134a (fully counterclockwise).

    d. Back flush the condenser and the liquid line by introducing flushing refrigerant into the supported end of the liquid line with the flushing nozzle. Hold the nozzle firmly against the open end of the liquid line.

    e. After the liquid line and condenser have been flushed, lay the charging cylinder on its side so R-12 or R-134a will not force more of the flushing refrigerant into the liquid line. Press the nozzle firmly to the liquid line and admit the R-12 or R-134a to force all of the flushing refrigerant from the liquid line and condenser.

    f. Remove the 1/2-inch hose and clamp from the condenser inlet connection.

g. Stand the flushing cylinder upright and flush the compressor discharge hose. Secure it so the flushing refrigerant goes into the waste container.

h. Close the dispensing valve of the R-12 or R-134a can (fully clockwise). If there is any flushing refrigerant in the cylinder, it may be left there until the next flushing job. Put the flushing kit and R-12 or R-134a can in a suitable storage location.

i. Install the new lubricated O rings on the spring lock coupling male fittings on both the condenser inlet and the liquid lines. Assemble the couplings.

9. Connect all refrigerant lines. All connections should be cleaned and new O rings should be used. Lubricate new O rings with clean refrigerant oil.

10. Connect a charging station or manifold gauge set and charge the system with 1 pound of R-12 or R-134a. (Do not evacuate the system until after it has been leak tested.)

11. Leak test all connections and components with a flame type of leak detector or an electronic leak detector. If no leaks are found, go to Step 12. If leaks are found, service as necessary; check the system and then go to Step 12.

12. Evacuate and charge the system with a specified amount of R-12 or R-134a. Operate the system to ensure it is cooling properly.

## SAFETY PRECAUTIONS

The use of safety when handling or using refrigerants can never be stressed enough. As discussed in chapter 6 of this TRAMAN, routinely think of safety for yourself and coworkers.

Extreme care must be taken to prevent any liquid refrigerant from coming in contact with the skin and especially the eyes. A bottle of sterile mineral oil and a quantity of weak boric acid solution must always be kept nearby when servicing the air-conditioning system. Should any liquid refrigerant get into your eyes, immediately use a few drops of mineral oil to wash them out; then wash the eyes clean with the weak boric acid solution. Seek a doctor's aid immediately even though irritation may have ceased. Always wear safety goggles when servicing any part of the refrigerant system.

To avoid a dangerous explosion, never weld, solder, steam clean, bake body finishes, or use any excessive amount of heat on or in the immediate area of any part of the refrigerant system or refrigerant supply tank, while they are closed to the atmosphere whether filled with refrigerant or not.

The liquid refrigerant evaporates so rapidly that the resulting refrigerant gas displaces the air surrounding the area where the refrigerant is released. To prevent possible suffocation in enclosed areas, always discharge the refrigerant into recycling/reclaiming equipment. Always maintain good ventilation surrounding the work area.

Although R-12 gas, under normal conditions, is nonpoisonous, the discharge of refrigerant gas near an open flame can produce a very poisonous gas. This gas also attacks all bright metal surfaces. This poisonous gas is generated when the flame type of leak detector is used. Avoid inhaling the fumes from the leak detector. Ensure that R-12 is both stored and installed according to all federal, state and local ordinances.

When admitting R-12 or R-134a gas into the cooling unit, always keep the tank in an upright position. If the tank is on its side or upside down, liquid R-12 or R-134 enters the system and may damage the compressor.

## TRUCK AND BUS AIR CONDITIONING

The cabs of many truck-tractors and long distance hauling trucks and earthmover cabs are air-conditioned. Most of this equipment is of the "hang on" type and is installed after the cab has been made.

Some truck air-conditioning units have two evaporators—one for the cab and one for the relief driver's quarters in back of the driver. Some systems use a remote condenser, mounted on the roof of the cab. This type of installation removes the condenser from in front of the radiator, so the radiator can operate at full efficiency. This is especially important during long pulls in low gear.

The system is similar to the automobile air conditioner and is installed and serviced in the same general way.

The air conditioning of buses has progressed rapidly. Because of the large size of the unit, most bus air-conditioning systems use a separate gasoline engine with an automatic starting device to drive the compressor. The system is standard in construction except for the condensing unit. It is made as compact as possible and generally is installed in the bus, so it can be easily reached for servicing.

Condensing units are often mounted on rails with flexible suction and liquid lines to permit sliding the condensing unit out of the bus body to aid in servicing.

Air-cooled condensers are used. Thermostatic expansion valve refrigerant controls are standard. Finned blower evaporators are also used.

The duct system usually runs between a false ceiling and the roof of the bus. The ducts, usually one on each side of the bus, have grilles at the passenger seats. The passengers may control the grille by opening and closing.

# CERTIFICATION

The Environmental Protection Agency (EPA) has established as per the Clean Air Act (CAA) that all technicians who maintain or repair air-conditioning or refrigeration equipment or technicians who operate recycling, reclaiming, and recovery equipment must be certified. Certification is administered by organizations with certification programs that are approved by the EPA. It is important to understand, that as a Utilitiesman, if you are not certified, you cannot do any HVAC/R service that requires use or removal of refrigerants. Certification requirements are divided into two different areas—automotive air-conditioning and HVAC/R.

## Automotive Air-Conditioning Certification

Automotive air conditioning is serviced or repaired more often than other types of air-conditioning systems. In today's world, automotive air-conditioning systems are heavily used as our society spends more and more time in their vehicles. Industry experts say that 25 percent of the R-12 purchased in the United States is used in automotive air conditioning. The fittings and hoses used in automotive air conditioning allow leakage to occur. Automotive air-conditioning service facilities or technicians are now changing (retrofitting) systems in vehicles to use refrigerant R-134a and removing CFC R-12 to meet new standards. From the EPA's standpoint, technicians must be meet the following requirements to be certified:

- Be aware that venting refrigerant is illegal.
- Understand why all the regulations are being created. Understand what is happening to the environment.
- Have a working knowledge of SAE standards J-1989, J-1990, and J-1991.
- Perform service in a safe manner without injuring personnel or damaging equipment. Areas that must be understood include venting, handling, transporting, and disposing of refrigerant.

Once these requirements are met through testing of the individual applicant, a certification card is issued.

## Heating, Ventilating, Air Conditioning, and Refrigeration Certification

Certification requirements to service standard types of air-conditioning systems are the same as for automotive air-conditioning certification. Unlike the automotive certification program, standard air-conditioning certification is divided into levels corresponding to the type of service the technician performs. There are four types of certification:

- Type I – Servicing small appliances
- Type II – Servicing high or very high-pressure appliances
- Type III – Servicing or disposing of low-pressure appliances
- Type IV (Universal) – Servicing all types of equipment

Individuals will be required to take a proctored, closed book test. These tests are offered by organizations approved by the EPA for the specific type of certification that the individual technician requires. Technicians can only work on air-conditioning systems that they have been certified for service.

Q33. The saturation temperature increases or decreases depending upon what factor?

Q34. What are the three basic types of automotive compressors?

Q35. A scotch-yoke compressor changes rotary motion into what type of motion?

Q36. Refrigerant can be put into a system when the service valve is back-seated. True/False

Q37. The POA valve, receiver-drier, expansion valve, and sight glass are combined in what type of device?

Q38. Service procedures for VIR systems are different than conventional automotive air-conditioning systems. True/False

Q39. What is the most important thing you should know before you perform a diagnosis on a system problem?

Q40. A refrigerant leak appears in what way at the point of the leak?

Q41. What is the most common cause of leaks on automotive air-conditioning systems?

Q42. For a refrigerant to be a suitable flushing agent, it must remain in what state during flushing operations?

Q43. Which part of an automotive air-conditioning system is replaced because it is impossible to clean?

Q44. A type IV certification is also known as what type of certification?

Q45. Who approves organizations to certify technicians?

# DUCTWORK

**Learning Objective:** Understand the basic types of ductwork systems and the components of those systems for distribution of conditioned air.

Distributed air must be clean, provide the proper amount of ventilation, and absorb enough heat to cool the conditioned spaces. To deliver air to the conditioned space, air carriers are required, which are called ducts. Ducts work on the principle of air pressure difference. If a pressure difference exists, air will flow from an area of high pressure to an area of low pressure. The larger this difference, the faster the air will flow to the low-pressure area.

## CLASSIFICATION OF DUCTS

There are three common classifications of ducts—conditioned air ducts, recirculating-air ducts, and fresh-air ducts. Conditioned air ducts carry conditioned air from the air conditioner and distribute it to the conditioned area. Recirculating air ducts take air from the conditioned space and distribute it back into the air conditioner system. Fresh air ducts bring fresh air into the air-conditioning system from outside the conditioned space.

Ducts commonly used for carrying air are of a round, square, or rectangular shape. The most efficient duct is a round duct, based on the volume of air handled per perimeter distance. In other words, less material is needed for the same capacity as a square or rectangular duct.

Square or rectangular duct fits better to building construction. It fits above ceilings and into walls and is much easier to install between joists and studs.

## TYPES OF DUCT SYSTEMS

There are several types of supply duct systems (fig. 7-42) that deliver air to room(s) and then return the air from the room(s) to the cooling (evaporator) system. These supply systems can be grouped into four types:

1. Individual round pipe system
2. Extended plenum system
3. Reducing trunk system
4. Combination (of two or more systems)

Return air systems are normally of three types—single return, multiple return (fig. 7-42), or combination of the two systems.

## CONSTRUCTION

Ducts may be made of metal, wood, ceramic, and plastic. Most commonly used is sheet steel coated with zinc (galvanized steel). Sheet metal brakes and forming machines are used in fabricating ducts. Elbows and other connections, such as branches, are designed using geometric principles. Some types of duct connections used in constructing duct systems are shown in figure 7-43.

Sheet metal ducts expand and contract as they heat and cool. Fabric joints are often used to absorb this movement. Fabric joints should also be used where the duct connects to the air conditioner. Many ducts are insulated to lower noise and reduce heat transfer. The insulation can be on the inside or the outside of the duct. Adhesives or metal clips are commonly used to fasten the insulation to the duct. As we are only briefly discussing construction here, you can find construction and fabrication methods in the *Steelworker*, volume 2. It details design and fabrication of steel ductwork.

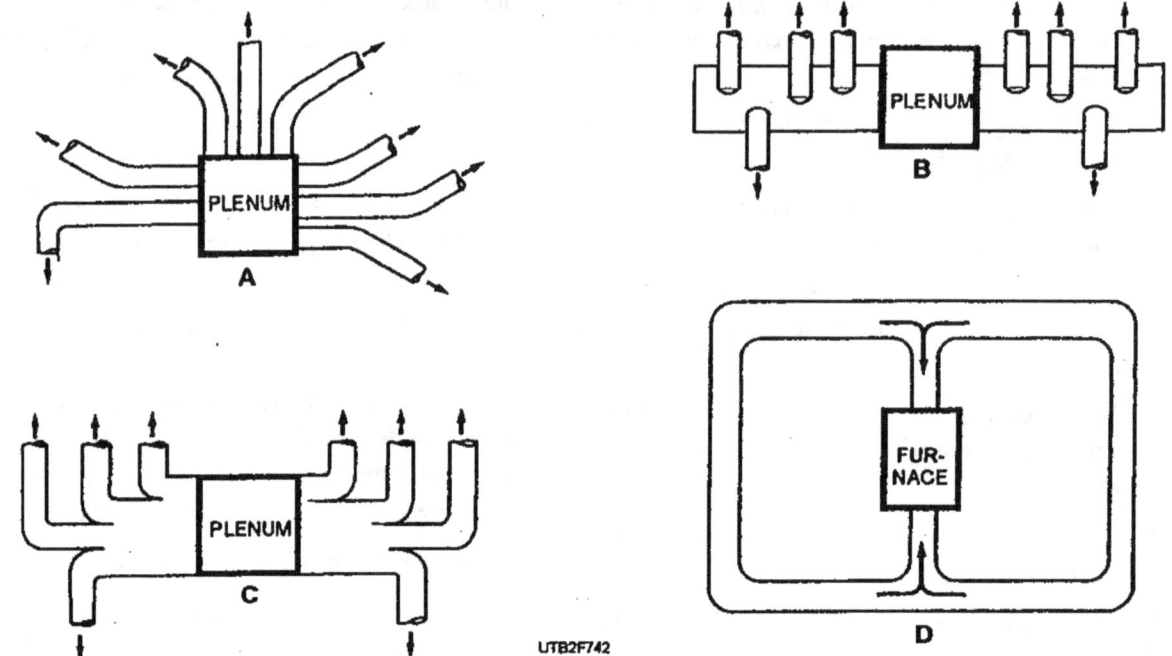

Figure 7-42.—Supply duct systems: A. Individual round pipe; B. Extended plenum; C. Reducing trunk; D. Multiple return air system.

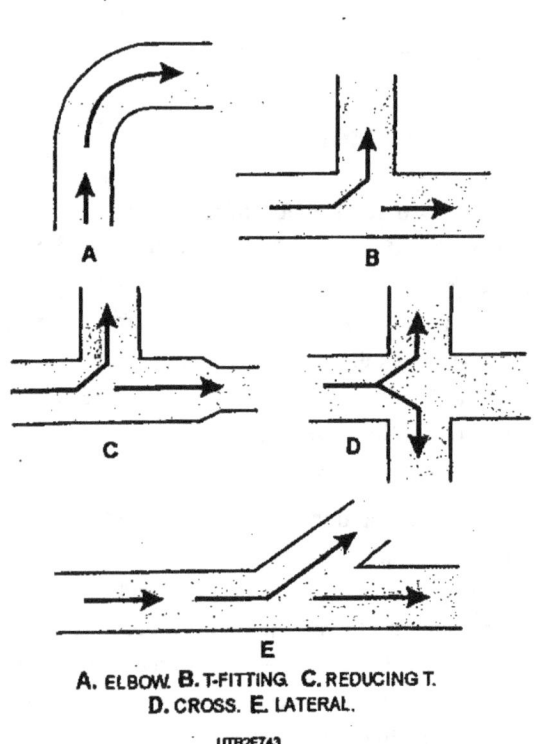

A. ELBOW. B. T-FITTING. C. REDUCING T.
D. CROSS. E. LATERAL.

Figure 7-43.—Typical duct connections: A. Elbow; B. Tee: C. Reducing tee; D. Cross: E. Lateral.

## COMPONENTS

To enable a duct system to circulate air at the proper velocity and volume to the proper conditioned areas, you can use different components within the duct system, such as diffusers, grilles, and dampers.

### Diffusers, Grilles, and Registers

Room openings to ducts have several devices that control the airflow and keep large objects out of the duct. These devices are called diffusers, grilles, and registers. Diffusers deliver fan-shaped airflow into a room. Duct air mixes with some room air in certain types of diffusers.

Grilles control the distance, height, spread of air-throw, and amount of air. Grilles cause some resistance to airflow. Grille cross-section pieces block about 30 percent of the air. Because of this reason and to reduce noise, cross sections are usually enlarged at the grille. Grilles have many different designs, such as fixed vanes which force air in one direction, or adjustable to force air in different directions.

Registers are used to deliver a concentrated air stream into a room, and many have one-way or two-way adjustable air stream deflectors.

## Dampers

One way of getting even air distribution is through the use of duct dampers. Dampers balance airflow or can shut off or open certain ducts for zone control. Some are located in the grille, and some are in the duct itself. There are three types of dampers used in air-conditioning ductwork—butterfly, multiple blade, and split damper (fig. 7-44). When installing a damper, always draw a line on temperature control.

## Fire Dampers

Automatic fire dampers should be installed in all vertical ducts. Ducts, especially vertical ducts, will carry fumes and flames from fires. Fire dampers must be inspected and tested at least once a year to be sure they are in proper working order. There are two types of fire dampers, which are fail-safe units—spring-loaded to close and weight-loaded to close. Fire dampers are usually held open by a fusible link. Heat will melt the link and the damper will close by either gravity, weights, or springs (fig. 7-45).

## Fans

Air movement is usually produced by some type of forced airflow. Fans are normally located in the inlet of the air conditioner. Air is moved by creating either a positive pressure or negative pressure in the ductwork. The two most popular types of fans are the axial flow (propeller) or radial flow (squirrel cage) (fig. 7-46).

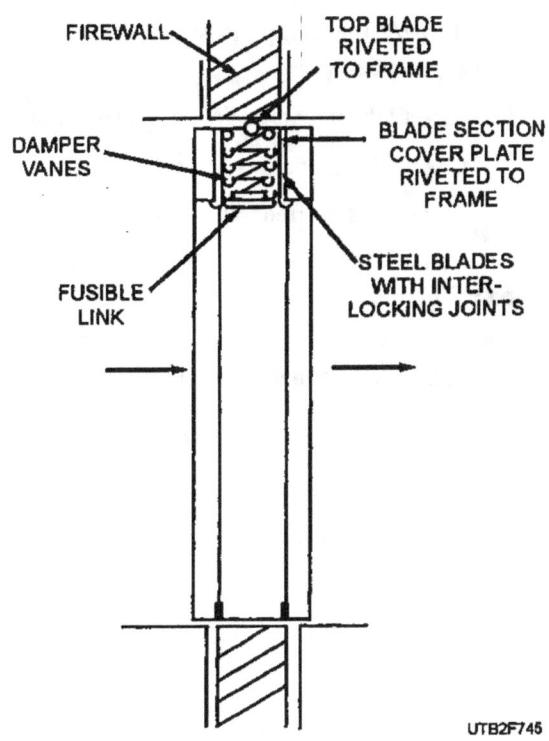

Figure 7-45.—Fire damper in OPEN position.

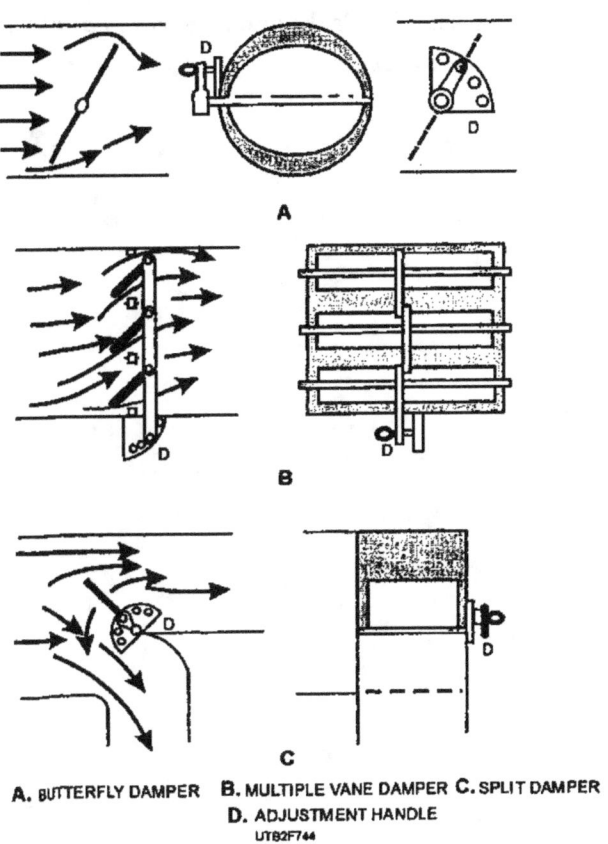

Figure 7-44.—Three types of duct dampers.

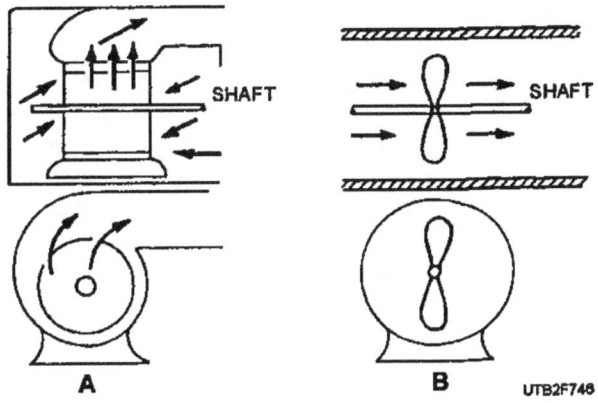

Figure 7-46.—Principal types of fans: A. Radial flow; B. Axial flow.

The axial-flow fan is usually direct-driven by mounting the fan blades on the motor shaft. The radial-flow fan is normally belt-driven but can also be direct-driven.

## BALANCING THE SYSTEM

Balancing a system basically means sizing the ducts and adjusting the dampers to ensure each room receives the correct amount of air. To balance a system, follow these steps:

1. Inspect the complete system; locate all ducts, openings, and dampers.

2. Open all dampers in the ducts and at the grilles.

3. Check the velocities at each outlet.

4. Measure the "free" grille area.

5. Calculate the volume at each outlet. Velocity x Area = Volume

6. Area in square inches divided by 144 multiplied by feet per minute equals cubic feet/minute.

7. Total the cubic feet/minute.

8. Determine the floor areas of each room. Add to determine total area.

9. Determine the cfm for each room. The area of the room divided by the total floor area multiplied by the total cfm equals cfm for the room.

10. Adjust duct dampers and grille dampers to obtain these values.

11. Recheck all outlet grilles.

In some cases, it may be necessary to overcome excess duct resistance by installing an air duct booster. These are fans used to increase airflow when a duct is too small, too long, or has too many elbows.

Q46. *What are the three common types of ducts?*

Q47. *What are the three types of return air systems?*

Q48. *Sheet metal ducts expand and contract as they heat and cool. True /False*

Q49. *What are the three types of dampers?*

Q50. *Once you have checked the velocities at each outlet, what is the next step when balancing the duct system?*